SUSTAINABILITY OF RICE PRODUCTION IN THAILAND

Sustainability of Rice Production in Thailand

Pascale M. Phélinas

Nova Science Publishers, Inc.
Huntington, New York

Senior Editors: Susan Boriotti and Donna Dennis
Office Manager: Annette Hellinger
Graphics: Wanda Serrano and Dorothy Marczak
Coordinating Editor: Tatiana Shohov
Book Production: Matthew Kozlowski, Jonathan Rose, Jennifer Vogt
Circulation: Cathy DeGregory and Ave Maria Gonzalez
Communications and Acquisition: Serge Shohov

Library of Congress Cataloging-in-Publication Data

Available upon Request

ISBN 1-59033-074-9

227 Main Street, Suite 100
Huntington, New York 11743
Tele. 631-424-6682 Fax 631-424-4666
e-mail: Novascience@earthlink.net
Web Site: http://www.nexusworld.com/nova

Printed in the United States of America

CONTENTS

INTRODUCTION

RICE IN THAILAND'S ECONOMIC HISTORY

All along the twentieth century, Thailand moved from a subsistence agriculture-based economy to commercial farming economy towards an increasingly industrialized economy. The process of industrialization led to the loss of agriculture's traditional primacy. Contributions from the agricultural sector to the GDP declined significantly from 35 percent in the 1960s to less than 10 percent by the end of 1999, while the manufacturing sector share rose to 35 percent. Exports of manufactured goods and non-factor services increased from a 25 percent share in GDP in 1980 to over 40 percent in 1999. However, despite the emergence of Thailand as a newly industrializing country, the share of agricultural employment in the Thai labor force remains close to 50 percent.

Up to 1995 Thailand was one of the fastest growing economies in Asia and the Pacific region; the economy flourished, recording an 8-8.5 percent per annum GDP growth rate. However, during the 1990s, many factors contributed to a gradual loss of competitiveness in labor-intensive products on international markets. Low world economic growth, increasing US, European and Japanese trade restrictions made Thai exports more difficult, whereas a high domestic inflation rate, high interest rates owing to the Bank of Thailand's tight monetary policy, real baht appreciation against many currencies and rising labor costs contributed to worsening the export picture. Meanwhile, the easing of foreign exchange controls generated a heavy influx of foreign capital, the majority in the form of short-term debt held by the private sector. Because the Central Bank maintained the basket peg system introduced in the mid-1980s, which, in practice, fixed the baht to the US dollar, Thai companies as well as financial institutions, seeing no exchange risks, largely borrowed money from international markets at interest rates lower than those available on the local market. Easing of lending criteria led to long-term unproductive investments in the real estate sector at highly inflated prices and to speculative activities on the stock market. The country's outstanding external debt rose from 33.8 percent of GDP in 1990 to 50.9 percent of GDP at the end of 1996, and the share of short-term debt to total foreign debt increased from 35.5 percent in 1990 to 55.6 percent in 1995.

All these factors resulted in the deepest economic downturn that the country ever experienced. Speculative attacks against the baht began after the Mexican crisis in 1994 and intensified in 1996 and in early 1997. On 2 July 1997 the monetary authorities introduced a managed float of the baht, which subsequently depreciated about 50 percent against the US dollar. The turmoil in the financial market that this depreciation brought about resulted in large capital outflows. Afterwards, a severe liquidity shortage occurred in the entire financial system, which, in turn, severely damaged the potentials of the real sector. The Thai economy further contracted by 0.4 percent in 1997 and 9,4 percent in 1998.

Since the explosion of the crisis, significant progress in short-term economic adjustment has occurred. The government succeeded in maintaining exchange and interest rate stability through tight monetary and fiscal policy, restoring its foreign reserve position so as to regain market confidence, restructuring and reforming the financial sector. But restoring the country's solvency has been achieved at very high cost in terms of economic contraction. Although after two years of tough economic conditions the economy finally began to show signs of recovery in 1999, many questions remain about the long-term sustainability of this recovery. Banks and finance companies remain under pressure from the high level of non-performing loans; the progress in debt restructuring is slow; the amount of public debt has sharply risen from 15 percent GDP in 1997 to 64 percent of GDP in 2000; export performance remains sluggish; private sector investment and consumption are still below pre-crisis levels; unemployment has reached about 1,7 million people and is not expected to ease as many firms are still downsizing their staff.

Rice has long been the most important food crop in the Thailand's agricultural economy. The Kingdom is the world's leading exporter of rice, selling abroad between four and six million tons per year with an annual average value of 40 billion baht, representing a world market share of about 30 percent. Although Thailand's trade figures have continued to show the economy's industrialization, the top ten Thai exports still include rice. Nowadays rice accounts for about 50 percent of the total area cultivated, 30 percent of the total value of agricultural production, and 12 percent of agricultural exports. Rice-growing households constitute 75 percent of the 5 million farming families, accounting for nearly 50 percent of the agricultural labor force. The performance of Thailand's rice sector is therefore of critical importance to the welfare of a large part of the country's population.

As for many other agricultural products, the rising costs of agricultural inputs resulting from fast economic growth coupled with the real baht appreciation had gradually weakened the competitiveness of Thai rice exports during the 1990s. Meanwhile competition from Vietnam, China and Pakistan in the low-quality rice category strengthened. Vietnam recorded its largest volume of rice exports in 1996 since it began exporting in 1989, and has now replaced the United States as the world's third largest rice exporter. As a result, the annual rate of growth of rice exports decreased from 8.3% in the 1980s to 5.4% during the 1990s. On a short-term horizon, rice exports have rather benefited from the crisis because the weakened baht contributed to supporting Thai rice competitiveness. Outward shipments of rice in 1997 were on par with 1996 exports (5.6 million tons); rice exports in 1998 reached a record of 6.4 million tons which was exceeded in 1999 since Thailand shipped about 6.8 million tons.

Until the 1980s, there was no doubt that Thailand had made the right choice in exploiting its potential of low-cost expanding of rice cultivation, which contributed to overall economic growth, foreign exchange earnings, and labor absorption. The radical transformation of rice production conditions that occurred with accelerated economic growth, as well as the progressive depletion of natural resources has presented a serious challenge for rice farming sustainability. The nation's land frontier has been reached and land resources have become increasingly scarce; the labor market tightening brought about by the changes in the production structure has resulted in shortages of labor; the amount of water available for paddy fields has diminished because the expansion in the supply of water has not been able to meet increasing demands as industrial and municipal uses of water have grown.

A shift to a new development strategy for rice has emerged as an important condition for continued development. Rice output expansion and export competitiveness strengthening had to be achieved using less land, less water, and less labor while at the same time improving the welfare of farmers on a sustainable and economical basis. In this context, the farm families increasingly had to recombine and improve the use of their productive resources of land, labor and capital both within and outside the farm business. The challenges that rice growers had to face include the following major issues, which are some of the central issues discussed in this book:

- Land scarcity
- Labor shortages
- Yield stagnation
- Difficult access to credit

After Open Access to Land, The Market?

Generally speaking, land surface is not only an essential input in the production function, but also an asset that provides a secure store of value and a good hedge against inflation. In agrarian economies it usually represents one of the most attractive investments for people with significant savings. Land is also valued for social, cultural or political motives because land ownership (or at least control of land) often enhances political power and prestige. Therefore, land rights distribution matters for both agricultural production and the distribution of an agrarian society's most valuable resource.

Even though historically Thailand has had plentiful land resources, continuous high population growth rates have resulted in rapid increases in rural labor population pressing hard on limited agricultural land. The outcome of this process is well known: a relatively low ratio of cultivated land per worker and small-sized individual farming units. The formation of this agrarian structure was further strengthened by the successive Thai governments which passed legislation so as to encourage small-scale producers and to promote a relatively egalitarian pattern of land ownership on the grounds that a myriad of small farms supply employment for the majority of rural dwellers and yield a more equitable distribution of income.

Nowadays, the prevalent view on land tenure is that starting from a historical base with farmers cultivating the land they owned, there has been a steady trend toward economic hardship, indebtedness and loss of land through the operation of the rural land markets. According to this view, land ownership concentration and its corresponding land scarcity has bred tenancy expansion and associated poverty. We will show evidence that questions the paradigm of large landowners exploiting landless and land-poor tenant. A quite different explanation for the expansion of tenancy in specific areas of rural Thailand is provided. Interesting new trends that appeared such as "capitalist tenancy" (large farms leasing in land) and reverse leasing (land-poor households leasing out land) led to new conclusions with respect to the rationale behind tenancy arrangements.

Another source of concern is the small and fragmented nature of rice holdings. Transforming tenure structure through the market mechanism could be an extremely helpful means of land consolidation. However, there is ample evidence available showing that land and capital markets in Thailand are characterized by multiple imperfections such as segmentation, asymmetric information, and high transaction costs. As a result, putting land on the sale market or purchase land is often an unattainable objective or a risky deal. In this context, tenancy arrangements through the land market provide a mechanism of land allocation, which can transform the pattern of ownership holding into a distribution of operational holdings which more closely correspond to an efficient distribution.

Increasing attention has been given to land titling in Thailand. Throughout the twentieth century, though lands were granted on a custom basis to provide a source of livelihood, the decision was, in many cases, not followed by appropriate arrangements of land rights. In 1982, only 12 percent of cultivated area was properly documented, while 49 percent was held under less secure forms, 18 percent by persons who lacked any officially recognized documentation, and 21 percent was illegally occupied forest reserve (Rattanabirasongse, 1998). Thailand therefore engaged in a large Land Titling Project in 1984 in order to provide the beneficiaries the advantages associated with full rights on the land owned. Greater tenure security was expected to raise farm productivity, boost the market value of land, and make access to credit easier. The rationale behind this policy, however, can be seriously questioned. In an environment of imperfect markets, small farmers access to credit is rationed, the cost of capital-intensive technologies they face differ from those relevant to large farmers, and a title to land may not overcome the obstacles associated with a small farm size.

Chapter one focuses on the changes in the tenure system observed in the rice subsector as a result of the closing land frontier. More specific questions are: has the structure of land ownership and landholding been greatly altered by the closing land frontier? Is tenancy still a minor concern in Thailand? To what extent have rural land markets developed? Have they contributed to the adjustment of landholdings to their optimal levels? Have land prices and land rents strongly appreciated, making access to land and/or land accumulation more expensive and thus less equitable? Has the strategy of land rights formalization ensured equity, productivity and sustainability?

Intersectoral Mobility and Labor Market Adjustment

Economic development requires change in resource allocation. As economies develop, the structure of employment in the economy usually shifts away relatively from agriculture to industry and services. In this process, agriculture is identified as the traditional sector that must draw surplus labor from occupations of low productivity to higher-productivity jobs in the industrial (and often urban-based) sector. This orthodox view of economic change assumes that people make complete transitions from one type of activity to another as development proceeds. Sectors are clearly distinguished and agriculture is contrasted with industry, rural is contrasted with urban. Much of the recent work, however, has shown that the development process, when viewed within the context of such dualistic models, can lead to misleading conclusions about the relationships between agricultural growth and industrialization.

Thailand has exhibited declining labor force participation in the agricultural sector since the 1960s and diversification of employment towards the non-agricultural sectors as a result of industrialization. Since the 1980s, in particular, non-agricultural employment has rapidly grown and the ability of non-agricultural sectors to absorb significant numbers of persons displaced from agriculture has correspondingly been consistent. Besides the size and persistence of these shifts, diversification of rural households' income over a wide range of economic activities has increased as well over time. Indeed,

Thai rice households generate significant shares of their incomes off the farm, usually through a multitude of short-term movements of workers between the rural and the urban sector. Growing evidence of importance of off-farm activities does not fit well into the conventional picture of dual labor markets nor supports the hypothesis of dissociation of urban and rural labor markets.

Two divergent interpretations of the process of diversification of employment within farm households have been suggested. They are broadly termed agricultural growth led diversification and distress diversification. The first hypothesis is that agricultural growth is the engine of income growth, which pulls farm expansion into non-agricultural activities, while the second hypothesis is that diversification of farms is induced by hardship in farming and by the need for rural households to find alternative sources of income. Diversification can therefore occur from a position of strength in which prosperous households invest the surplus generated from the primary activity into non-farm activities. Multiple economic activities can also occur from a position of vulnerability and denote a strategy of survival in the context of highly seasonality in agriculture, poor harvests, uncertainty and risks in production. A better understanding of diversification therefore implies careful consideration of the strategies by which rural families allocate the working time of their members.

Most of the existing literature on labor absorption in agriculture puts forward the skewed distribution of land toward small farms and the associated predominance of labor-intensive techniques as the main explanation for the patterns of labor use. The existence of an inverse relationship between farm size and labor intensity constitutes one of the issues widely discussed among economists. Imperfections in the labor market have generally been put forward to explain why a larger family labor input is associated with smaller farms. Imperfect labor market, however, cannot alone account for this relationship; it has to be clarified why farmers fail to adjust labor endowments through land rental market or through sales of fixed factors. We provide empirical evidence showing that this inverse relationship reflects multiple market failures not only in the market for labor, but also in the market for land and credit.

The recent context of labor-scarce economy has put rural labor market under strong pressure and raised doubts about the ability of rice farms to remain competitive unless labor-saving technologies are widely adopted. Allowing labor intensity and overall labor use in rice farming to decrease is therefore one of the crucial problems the rice sector has tried to address. The intensity of use of capital inputs is a key determinant of the labor regime under which rice is produced. However, it is difficult to assess ex ante the impact of technological change on employment. The introduction of machines in farming systems is generally assumed to substantially reduce the need for labor. However, mechanization can be selectively concentrated in operations where labor displacement is not the primary effect. In addition, mechanization can be accompanied by the introduction of labor-using technologies so that the saving of labor days due to mechanization can be more than compensated by higher cropping intensities and higher labor application to crop care, harvesting and threshing activities. The question about whether substantial savings in labor employment in rice farming occurred with concomitant increases in capital formation is therefore an empirical one.

In chapter two, the factors governing the labor demand and supply conditions in rice farming are analyzed with empirical data. Three important themes are addressed: first, investigating the factors causing the observed differences in labor use between farms of different sizes. Second, evaluating the impact of technological change on labor absorption in rice farming. Third, providing information on the nature and size of Thai rice farmers earnings from off farm activities and drawing inferences concerning the rural labor market functioning.

INVESTMENT IN RICE FARMING: TOWARDS INTENSIFICATION OR TOWARDS MECHANIZATION?

From a macroeconomic perspective, in an economy where many employment opportunities are being created in the agricultural sector, rapid and efficient increase of agricultural output is essential to meeting overall growth goals and strengthening economic development. Studies on the historical experience of developed countries have stressed the role of technical change in enabling agriculture to make a significant contribution to development. In these countries, technologically induced increases in the productivity of the existing on-farm resources of labor and land were of decisive importance in enlarging foreign exchange earnings and made possible the simultaneous growth of industry and agriculture. In agrarian economies the pace and pattern of agricultural output growth has been strongly affected by the types of technical innovations that were introduced. Therefore, understanding the many factors that influence the emergence and adoption of innovations is one of the keys to understanding output growth and its potential.

There is no generally accepted theory of innovation and the process of technological adoption in agriculture is not yet well understood. There have been two main theoretical approaches to understanding the sources of technical change in agriculture. On one hand, Boserup (1965) has showed the existence of a positive correlation between the intensity of cultivation practices and population density. On the other hand, the induced innovation hypothesis has tried to explain why a particular technology occurs in terms of the relative scarcity of resources which guides technological progress toward saving the factor that has become relatively expensive (Binswanger and Ruttan, 1978; Hayami and Ruttan, 1970). It has also been observed that the successes of the technological packages were closely linked to the existence of favorable institutional environments. In particular, lack of public investment in irrigation infrastructure, education, extension, communication network, as well as low density in marketing and banking infrastructure have generally acted as disincentives to the adoption of more productive farming systems.

Thailand's relatively high growth rate of rice output has traditionally come from the expansion of cultivation frontiers to more and more marginal areas. For a long time, rice farmers have pursued extensification strategies in response to population growth and increased exports demand. This extensification path became virtually unsustainable during the 1980s as the Kingdom experienced a succession of mutually reinforcing shifts in its factor endowments. First, most land suitable for rice farming became exhausted. Second, the accelerated switch in the composition of the national output that occurred had wide repercussions on labor scarcity and labor use intensity which challenged past practices of rice farming. Third, demand for water escalated, contributing to intensified competition among users. Increased factor scarcities drove factor prices up or resulted in quantity rationing, which both created the conditions for capital-using intensification.

Further output growth can rely on a widespread adoption of green revolution technology. However, the adoption of such a package depends to a large extend on good water control to create conditions in which high-yielding modern varieties can thrive. Water availability and water use efficiency have been some of the many problems relating to low reliance on HYVs and thus yield stagnation in Thailand. Despite significant public efforts to develop irrigation systems, rainfed areas represent nearly 80 percent of the total rice area cultivated. Irrigation development has slowed in recent years because high costs and environmental problems have discouraged investment. In addition, disappointing performances of many large-scale irrigation schemes due to lack of water in the country's dams, typical unproductive water losses during conduction and water application on the farm, and inefficient management of water, have reopened the issue of the role of irrigated agriculture. The scope for further growth in grain production by using more land-saving inputs may thus be limited.

By contrast, mechanization has a dramatic potential to increase labor productivity. Sustainable rice production development depends heavily on agricultural engineering in areas where shortage of labor has become an important constraint to increased rice production. Whether operation specific or general labor saving machinery has been and/or will be adopted depends on many factors. Mechanization is generally attractive when it replaces repetitive, high energy-consuming human labor. The shift from draft animals to tractors is, in such cases, mainly determined by the relative costs of fodder and diesel, the reliability and cost of access to spare parts and servicing facilities. In an economic context characterized by labor scarcity, the main motivation in substituting machines for human labor is, accordingly, a rising rural wage. Since the intensity of use of labor varies according to farm size and farm location, predictions about which level of technology will be introduced and applied can thus only be made by careful analysis of each recipient situation.

The objective of Chapter three is to understand the pace of implementation of new techniques in the rice-farming sector. We show that the incentives for agricultural innovation have been very different depending on the farm characteristics. Farmers' investment decisions were critically dependent on factors such as local natural resources endowment, scale of operation, tenure security, access to capital market, and structure of relative prices of rice according to quality. Variations in the amount of public investments in roads, marketing facilities, irrigation services, were also crucial in the transition process.

The Challenge of Agricultural Finance

It has been a long-held belief that most rural households in developing countries lack access to adequate financial services for efficient intertemporal transfer of resources as well as for risk coping. Indeed, rural households engaged in agricultural activities face a multitude of risks affecting income generation and consumption. Furthermore, the seasonality in agriculture results in a payment system that often involves deferred or advance payments. There is a long time lag between the time at which expenditures are incurred for farm inputs and the time at which crop is harvested and sold. In such a context, providing a buffer for consumption shortfalls or income fluctuation through credit (or insurance) markets would stimulate farm investment by allowing households to take up higher-risk and higher-return activities and could result in considerable welfare gains for households.

Based on the assumption that most farmers were either too poor to save or temporarily liquidity constrained, accessible credit has been thought in many developing countries as a means to smooth farmers' consumption needs and hence boost farmers' investment. The government of Thailand is among a number of governments in developing countries, which have intervened actively in the agricultural credit market. In 1975, the Bank for Agriculture and Agricultural Cooperatives as well as commercial banks were enjoined to grant targeted lending to the rural sector at interest rates typically lower than the rates of alternative sources of funds. This policy resulted in sizeable increase of institutional lending to agriculture.

Even though expansion of institutional credit was implemented as a policy to increase access of a broad part of the population to cheap financial services, it is generally believed that the Thai government's policy of a fixed and low interest rate has resulted in a considerable rationing of formal credit, with a great number of rural households being denied access to such credit despite their demand for it. Private credit markets in Thailand remain active and the institutional market with uniform and relatively low rates of interest coexists with the so-called informal market that charges a widely dispersed set of relatively high rates.

Theoretically, credit market segmentation has been explained as arising from the presence of asymmetric information in the market and differences in the costs of screening, monitoring and contract enforcement across lenders (Stiglitz and Weiss, 1983). The absence of complete information about the contracting parties leads to special problems of moral hazard and adverse selection. As a result, interest rates do not play the simple allocative role ascribed by economic theory. Differences in interest rates reflect not only differences in cost of funds, transaction costs and risks but also the fact that funds and information do not flow freely between segments. The resulting credit market equilibrium can then be characterized by discrimination against small borrowers and credit rationing. In such a context, the informal sector usually serves those borrowers who are not eligible to bank credit and/or those who are rationed out by banks.

In theory, constrained liquidity adversely affects the utilization of cash inputs, dictates factor ratios that deviate from optimum, and ultimately leads to lower output. Where technical change open up profitable new opportunities for investment, credit bottlenecks may act as real constraints on the growth of agricultural output. Better access to finance as well as improved terms and conditions of financial contracts were thus expected to spur agricultural development through their influence on the adoption of new technologies. However, there has been little empirical evidence that rice production has been effectively constrained by lack of access to formal or government-controlled credit. The impact of credit programs on the adoption rate of new technologies in the rice sector thus remains an open question.

Chapter 4 reviews the evidence about the structure of the rural financial markets in Thailand. The roles that the formal and informal sector plays, in conjunction or in exclusion of each other, deserves close attention. We also shed some light on the determinants of the demand and supply of rural credit, and describe characteristics of rice households who succeeded in obtaining loans from various sources. Whether credit has played an essential role in the propagation of new technology among rice farmers is another central interrogation addressed in this chapter. The focus of government on credit as a primary instrument of agricultural development is analyzed as well.

SURVEY DESIGN AND SURVEY AREA

The data set used in this book was obtained from a unique survey carried out between November 1993 and June 1994, covering one agricultural year and two growing seasons; it encompasses 300 households distributed among

15 villages located in three provinces in Thailand: Suphan Buri in Central Plain, Pitchit in the North and Roi Et in the Northeast.

The sampling design of the survey proceeded from the following guideline. Because of the diversity in economic as well as in natural environment and, accordingly, the variety of constraints farm households face on their decision-making, we tried to locate our sample in agro-economic zones that exhibit significantly different environments for rice agriculture. Considering the information regarding cropped area under paddy in each region, cultivation practices, irrigation facilities and rainfall, density of communication network, proximity to Bangkok, three provinces were selected. Each of the provinces was deemed typical of the region in which it is located, and represent the major soil types, rainfall, cropping patterns, and economic conditions of Thai rice growing areas. Limited resources available have been a constraint, which required restraining the spatial scope of the sample. As a result, while representative of the major types of cultivation, the sample is not representative in a broader sense.

Suphan Buri is located in Central Plain, which is a flat alluvial area lying on the Chao Praya river delta. It has the most favorable watering conditions both because of heavier rains and of past public investments in irrigation facilities. Soils are clay with a high level of fertility but poor drainage. It is the most intensively cultivated area, commonly considered as the rice bowl of Thailand. Vegetables, fruit trees and livestock are other major activities carried out by farmers. In addition, Suphan Buri is undoubtedly privileged with access to roads, railways and river transport. Villages are thus well connected to final markets, a fact which makes this area highly commercialized.

The lower North, where Pitchit is located, can be considered as an extension of the Central region in terms of topography and soil fertility. However, water conditions are less propitious than in Central Plain. The extent of crop diversification has been great. Major agricultural products other than rice include maize, cassava, soybeans and garlic, fruit trees, vegetables and livestock. The province enjoy a quite good communication network.

Roi Et is located in the Northeast region, which is the most poorly endowed in terms of soil fertility and water control. It consists of a sandstone arid plateau where soil erosion and periodic droughts are acute. The dominant soil is loam and sandy loam characterized by low fertility and low ability to retain water. Diversification has occurred only into livestock. Low productivity, low-income levels further characterize Roi Et's economic environment. The province is located far away from main commercial and industrial centers and suffers from relative scarcity of communication facilities.

Districts and villages were randomly selected from a list of villages supplied by the provincial Department of Agriculture. Five of the villages are located in Suphan Buri, five are located in Pitchit, and five in Roi Et. The sample of farm households was then selected by random procedure using lists of households in each village. The total sample divides into approximately equal sized sub-samples of 100 farm households in each of the selected provinces.

A detailed household questionnaire was designed and pre-tested. Students of Chulalongkorn University, under my supervision, performed information collection. The survey contains about 25, 000 observations on land ownership, land rights, tenancy, land transactions since household formation, farming techniques (cropping intensity, nature of seeds, fertilizer use, ownership and use of machinery), households demographics, family members' labor input both on and off farm, hired or exchanged labor, incomes, valuation of assets, involvement in formal and informal credit markets. In addition, my research assistant and I conducted in-depth informal interviews with numerous villagers focusing on a wide range of topics such as perceived access to credit from the formal and informal sector, availability of water and quality of the irrigation system (if any), the rationale underlying the choice of the varieties cropped, the perceived source for achievement of "sub-optimum" yields, perceptions regarding scarcity of labor.

Chapter 1

ACCESS TO LAND: PATTERNS OF LAND HOLDING AND LAND OWNERSHIP

1.0 INTRODUCTION

Land resources have long been a key factor in Thailand's agricultural development. Until the 1980s the growth of agricultural production and exports had been dominated by a massive expansion in the land area under cultivation, giving the country a strong comparative advantage in agriculture. Since the 1980s however, Thailand is in an intermediate phase in which there is little scope for extending the area under cultivation because the land frontier has been reached. As a result, the costs of providing land have risen and certain population groups find access to land increasingly difficult.

The income of rural households depends heavily on the equity and efficiency with which land is distributed and used. Disparities in land holdings usually coincide with disparities in income. Access to factors of production other than land is largely determined by the size of land holdings and families with larger landholding tend to own more non-land resources. In particular, the ownership of stock of capital tends to be unequally distributed and to accentuate the initial unequal distribution of land. From the point of view of equity, the question of how the limited land is distributed among different groups of the population, who has the right to cultivate land and on what terms, is therefore a crucial one.

Land rights distribution is also a crucial question from the point of view of efficiency. The influence of agrarian structure upon agricultural productivity, technical change and employment in developing countries remains an issue of considerable controversy. Farmers' relations to the land, its control and use are believed to be essential factors of their behavior. Agrarian structure and land tenure have an influence on many decisions on farm investment, production techniques, and farm productivity. The view that dynamic agricultural development must depend upon large and modern farms often conflicts with the evidence that a small, family farm structure could be more favorable for agricultural productivity.

Land tenure is thought to influence agricultural productivity through the security effect. Because the establishment of clear ownership rights lowers the cost and risk of transferring land, land titling is expected to improve factor mobility, resource allocation, and thus, productivity. Dynamic effects of secure land tenure on the performance of farmers should result in increased willingness and ability to invest on the part of the secure farmers, because they are assured of reaping the future benefits of their present efforts. Moreover, when private land rights are well-established and legally protected and when they can be freely exchanged, land acquires collateral value and the supply of credit tends to increase. Investment is thus expected to be negatively related to the level of uncertainty regarding tenure.

Land being a decisive factor, this chapter considers the conditions which have determined access to land for the rice peasant producer, in particular the major part played by the land frontier on these conditions. The history of the land frontier and the changing of land rights over time are briefly reviewed in section one. The crucial role played by the state in setting the conditions that have determined access to land will be highlighted. Section two examines the patterns of land ownership among rice farmers. It assesses some commonly-held opinions about these patterns by examining the concentration of land ownership, the occurrence of land transfers and land prices. Section three discusses the role of land rental markets and the nature of landlord-tenant relationships.

1.1 The Closing Land Frontier and the Changing of Land Rights

In a predominantly agrarian economy, initial demographic conditions, the rate of population growth as well as the endowment of land resources have a very powerful effect on the entire course and pattern of development. In particular, population growth does not imply the lowering of labor to land ratios if the land frontier can be expanded or if labor absorption in the non-agricultural sector is fast. Actually, both options have been successively available to Thailand. The Thai history of the land frontier (and thus rice production) can be divided into three phases which are further detailed. The first phase lasted from 1855 to 1945 and was characterized by a huge expansion of cultivated area. Population growth was one cause of this expansion, but an equal or more important factor was the expansion of market production. Also, because population density on arable land increased, land tenure systems became increasingly exclusive. During the second phase which began at the end of the Second World War, the growth of the area cultivated slowed down while population continued to grow. This led to a slight decrease in labor/land ratios. The third phase started around the 1980s with the strong drop in the growth of cultivated area resulting from both the closing land frontier and the rapid expansion of the non-agricultural sector.

1.1.1 Population Pressure and Shifts in Land Use Patterns

Thailand has a long history of agricultural land expansion. The opening of the country to international trade in the 1850s triggered a long movement of expansion of rice cultivation. In the years that followed the Bowring treaty of 1855, Thailand specialized in the export of rice, which was the principal crop grown. Before this Treaty, farmers, who produced most basic goods (farm equipment, housing, clothing, etc.) had little reason to expand rice production much beyond their own needs. The demand for Thailand's rice in neighboring countries was erratic and the King might at any time limit or forbid the exportation of rice.

After 1855, both supply and demand factors accounted for the growth of rice trade and, consequently, expansion of rice cultivation. On the supply side, the new free trade policy gave farmers incentives to produce a marketable surplus of rice in exchange for imported commodities, and displaced native handicrafts production and cottage industries. On the demand side, there were a number of factors which created a regular and increasing foreign demand for Thai rice: modern transportation coupled with a decrease in long-distance transport costs, the development of mining and plantation production in Southeast Asian countries which then depended on large imports of rice, and population growth in Far-eastern countries like China (Ingram, 1971; Manarungsan, 1989).

Much land, but few people, was a chronic condition of the Thai state up to the last years of the nineteenth century. During this time the economy was characterized by a high land/man ratio. The control of manpower formed the basis of economic, political and social power (Feeny, 1982; Piker, 1976). Partly in response to the labor needs of the Kingdom, different classes of citizens were required to provide different amounts of *corvée* labor to the King or to their *patron*, and slavery was common. Many rural Thai entered bondage voluntarily, often to discharge debts. Freeman and slaves were tied to their overlords, not to the land, and as such, were exchanged or offered as gifts among nobles (Lasker, 1950). Commoners could shift their allegiance from one lord to another, a fact which encouraged a degree of geographical mobility on the part of the population. Migration, both forced and voluntary, was widespread and contributed to the lack of attachment of rural Thai to any particular plot of land (Mehl, 1986).

Thailand's increasing participation in international markets placed a premium on expansion of rice production. It was recognized that the labor increment required to achieve this goal could be realized only by the relaxation of the traditional labor obligations such as *corvée* as well as the abolition of slavery. Accordingly, human bondage was gradually but systematically abolished during the last decades of the nineteenth century and the state's labor requirements, previously met by the *corvée* labor system, were fulfilled by hired labor against wage payment. The release of the labor force from the rigid systems of *corvée* and slavery enabled the native Thai to exploit market opportunities.

The magnitude of the transformations which came in the wake of these developments is indicated in Table 1.1. The peasantry of Thailand, freed from traditional labor obligations, responded to income incentives by bringing new land everywhere under cultivation: the total area of paddy under cultivation grew at roughly 2.7%[1] per year between 1907 and 1955, while the number of rice workers increased at a rate of 2.1%[2]. The amount of cultivated area per

---

1 There are no national estimates for the land brought under cultivation for the years prior to 1905. Although the area crop series from 1905 until 1955 are subject to caution, the overall trend is basically reliable.

2 Labor force patterns for Thailand before the Second World War are at best poorly known. Estimates of rice labor force have thus employed a considerable level of *guestimation*. Our

rice worker therefore rose by 0.34%. Area expansion first took place in the relatively more accessible lands, particularly those situated around the capital, where the existing network of canals and streams provided convenient and cheap transportation means. By the quarter mark of the twentieth century, virtually all unused land in the Central Plain had been cleared and brought under cultivation. Only later, in the closing years of the 1950s, with the government sponsored creation of the railway network and the improvement of internal transportation which resulted in a reduction of marketing costs, was the opening of the North and Northeast to rice cultivation facilitated.

Table 1.1: Population Pressure, Planted Area Expansion and Forest Depletion

	Annual growth rates				
	Paddy production	Planted area	Number of workers	Planted area per worker	Forest area
1907-1955	1.73	2.71	2.07	0.34	
1955-1983	2.48	2.15	2.67	-1.03	-2.62
1983-1998	1.31	-1.00	0.79	-0.61	-1.51

SOURCE: Feeny (1982); Office of Agricultural Economics; Labor Force Surveys.

From the end of the Second World War up to the 1980s, the rate of increase in paddy land under cultivation slowed slightly down to 2.2% per year while the area planted per worker decreased by 1.03%. This contraction resulted mainly from strong population growth. Regionwise, the extension of rice cultivation proceeded considerably faster in the outer provinces than in the Central Plain. Part of the increase was required merely to feed the growing population of these regions. Although 53% of the rice production came from the older areas of settlement until the mid 1950s, new areas tended to contribute to national production at a higher rate: in the 1980s, the Northeast accounted for 38% of total production compared to 6% in the 1950s, while the North accounted for 30% compared to less than 2% in the 1950s. This does not mean that old lands were not profitably farmed, but that the relative mix of factors still favored extensive farming via incorporation of frontier areas rather than more intensive farming of already occupied, centrally located lands.

---

estimates are based on various population and agricultural censuses, and more recently on the

After years of rapid expansion up to the end of the 1970s, the supply of land for rice began to decrease slightly in absolute terms at a rate close to 1% per year during the 1980s and 1990s while the labor force in rice farming grew at a relatively modest rate of 0.8% per year. Meanwhile, domestic production of rice increased by only 1.3% compared to 2.5% from 1955 to 1980. These significant decelerating trends can be attributed to four main factors. First, the strong influence of export markets induced a process of diversification of agricultural production which led to the decline in the dominance of paddy cultivation in Thai agriculture relative to other crops (maize, sugar, kenaf, cassava) or products (fruits and vegetables, livestock, shrimp). Second, population growth drastically fell to 1.5% per year in the 1970s; because population growth translates into growth in labor force with a given time-lag, the impact on the labor force was felt in the mid-1980s. Third, the rapid development of the manufacturing and services sector led to competition for the use of the country's scarce resources. Since the 1980s, the growth of rice output has been associated with rapid expansion in nonfarm employment. The absorption of labor in the non-agricultural sector coupled with the continued shift from rice to other agricultural activities caused the rice labor force growth to slow down. Fourth, vacant land suitable for rice was no longer available and the land frontier could no longer be extended. As a result, the amount of cultivated land per person continued to decrease but at a much lower rate than through the prior period.

1.1.2 The Dramatic Depletion of Forest Land

Because most of the area currently under cultivation in Thailand was once covered by forests, especially outside the lower Central Plain, the main factor that has permitted the dramatic conversion of virgin land to agricultural use has been the depletion of forest land. Slash and burn cultivation was often practiced by new settlers to clear the land in order to make it suitable for paddy cultivation. From an area of 240.6 million rai (75% of total land area) in the early 1900s, the forest has now been reduced to an area of about 80 million rai (27% of total land area), despite repeated measures taken to protect the forest. Thailand thus lost about 65% of its forests.

Labor Force Surveys.

The extent to which rice cultivation is responsible for this phenomenon is quite difficult to assess based on aggregate figures. The rapid expansion of area under paddy cultivation in the pre-World War II period was concentrated in the lowland areas that were not heavily forested. In contrast, much of the expansion in the post-World War II period took place in formerly wooded areas[3]. Upon examination of different regions, the deforestation rate in the Northeast has been the highest, partly because of a faster rate of agricultural expansion resulting from low agricultural productivity per unit of area.

Forest encroachment by cultivators has occupied a central position in the debate on the deforestation of Thailand. Several factors can be put forward to explain this long movement of forest depletion. First of all, surface area under rice cultivation and its expansion into virgin forest has been a result of the relative profitability of frontier farming. Hence, any change in the variables which increased the profitability of frontier agriculture tended to augment deforestation. Given that people were free to move and open new land, all forest land with a positive land rent was cleared and transformed into agricultural production. For instance, up to the end of the nineteenth century, land taxes per hectare were reduced and new land was exempted from taxes in the first year so that land tax rates in Thailand were much lower than in neighboring countries. But most important of all, the strong population growth which occurred until the mid-1970s lowered the opportunity cost of labor, which, in turn, made the forest conversion profitable and enabled frontier agriculture to act as an employment residual.

Secondly, the location disadvantage of remote forest land for agricultural use has been reduced through the improvement of infrastructure such as railways, road networks and electricity. During the second half of the nineteenth century nine major canals were excavated and linked to the more distant areas of rice cultivation. This was followed by a substantial expansion of the railway network and improvement in roads during the first half of the twentieth century. When the Second World War came, 3130 kilometers of railway line were opened including lines to the upper North and the far Northeast (Ingram, 1971). A large share of the forest loss can thus be assigned to state promoted agricultural expansion facilitated by large scale transportation infrastructure construction.

---

3 Nevertheless, a high proportion of deforestation probably occurred through the expansion of upland crops.

Thirdly, forest clearance has also depended on how well the land rent could be captured under existing property rights arrangements. From this perspective land clearance can be viewed as a way to obtain and secure land rights, both according to customary and national law. At the turn of this century, land was still relatively abundant and the common practice to acquire land for cultivation purposes was carried out by settling open areas that were mostly forest areas. Clearing forest land as a way to claim land for cultivation area was an old traditional practice. This pattern of rural land acquisition still existed even after the promulgation of the Land Code of 1954.

Fourthly, the massive depletion of forest land has been attributed to the lack of coherent land policies as well as to a lack of management and enforcement. The Thai forestry policy originated in 1896 with the creation of the Royal Forest Department and the passage of the Forest Protection Act in 1897. Although conservation issues were discussed as early as 1910s, the main focus was placed on the regulation of commercial use of major forest products and on the issuance of free permits for cutting trees for domestic use until the 1930s (Onchan, 1990). Since that period, legislation has continued to reinforce the government's right to preserve forests mainly through the setting up of forest reserves. Virtually all forests have been proclaimed *de jure* state property and are officially reserved areas, publicly owned with their commercial use theoretically subjected to the Royal Forest Department authorization.

Though encroachment of forest land was a punishable offense under the Land Code, and even though the government took a clear position against encroachment, the government's attitude has never been consistent. It has been characterized by ambiguity, indifference, laxity and hence, inefficiency. Although forest laws had been enacted since 1897, there were no clear identification of reserved forest areas and other types of land use until 1932. After 1932 some areas were legally declared as forest reserves after the forests had been cleared and settled by villagers. As a result, reserve forests often overlapped with village farmland. Encroachments were also often accommodated after the fact by the local administration that issued special land documents for the newly settled areas. Continued deforestation resulted in an official lowered amount of forest land to be protected in reserves from 50% of the total area in the early 1950s to 40% in the National Forest Policy enacted in 1985 and down to a low of 25% at present. This policy, which has clearly given *de facto* private property rights to the land, is likely to continue:

in April 1997, the Forestry Department adopted a resolution which stipulated that land in conservation forests currently occupied by villagers would be downgraded to allow people to work the land under existing laws.

Ultimately, an insufficient understanding of the mechanics of deforestation and factors influencing the decision-making process of farmers explains why most attempts by the Thai government to control deforestation have not been very successful. Past forest policies were drafted without considering socioeconomic conditions and demographic trends. Increased crop planting and expansion into primary forest have been a response to increased crop profitability relative to alternative employment opportunities, expected land scarcity, and a race for property rights. Most farmers encroached on forest reserves in search of a better livelihood or as a way to avoid the declining productivity from their current farm holdings. Deforestation has thus been mainly caused by the limited sources of income other than agriculture.

From this last point of view, it is interesting to stress that, according to the Royal Forestry Department, the annual rate of forest encroachment has slowed down since the 1980s to 2.9 million rai per year compared to 4.5 millions previously. On one hand, this has certainly been a result of the lack of fertile forest land to clear. The supply of good land has been exhausted; land is increasingly hard to find and more costly to clear. Low accessibility and remoteness of remaining land also incur higher transportation costs and lower farmgate prices. On the other hand, industrialization and the resulting migration of workers from rural areas to urban centers, has undoubtedly reduced pressure on forests. Therefore, part of the solution to the problem of forest depletion is to be found outside the agricultural sector and would entail providing attractive alternative employment opportunities.

1.1.3 Land Rights and Ownership Security

The conventional wisdom of the evolutionary theory of land rights is that under the joint impact of increasing population pressure, advances in farming technology and the emergence of agricultural markets, land rights spontaneously evolve towards rising individualization through the pressure of land users for the creation of formalized private property rights. As long as land is abundant, the absence of property rights with their standard attributes of exclusivity and free transferability do no imply any uncertainty regarding the ability to utilize the land. However, when there is growing competition for the use of land, there tend to be an increased number of disputes over ownership of land, inheritance and land boundaries. This generally lead to the emergence of new property rights.

Historically, all land belonged to the King who used to grant long-term, usually non-transferable land leases to individuals on the condition that land was brought under cultivation. Any freeman had the right to claim some land according to a system of ranks called *Sakdi Na*. A common freeman had a *Sakdi Na* of 15 rai, that of titled officials ranged from 400 to 10,000 rai, while for royal persons it went as high as 100,000 rai (Kemp, 1981). Virtually all land was held on a right of occupancy, either on a statutory or granted right of occupancy. Although the rights of farmers were restricted to usufruct, in practice, *de facto* ownership emerged. Land was transmitted from generation to generation in the families of freemen and could also be sold by freemen as private property (Piker, 1976).

Until the first half of the nineteenth century, land was plentiful and rights to use land not contested. Accordingly, there was no formal testimony of land rights. During the second half of the nineteenth century, land rights began to evolve. This process was initiated by population pressure, the opening of the country to international trade and the increased commercialization of rice production. The resulting changes in factor endowments, reflected in price variations (in this case rising land prices) created gaps between the marginal returns and the marginal costs of factor inputs, which, in turn, gave rise to formal registration of private land rights and land titling procedures (Feeny, 1982; Feder, 1988).

Title documents began to be issued for rice land through the 1892 Land Law and then continued with the enforcement of the 1901 and 1936 land laws. Meanwhile, slavery was abolished. During this period, most of the titling effort was concentrated in the Central Plain, the main rice producing area. Later, the Land Code of 1954 defined the details for the application of land rights and contained procedures for the issuance of documents. The Land Code recognized four types of private ownership documents. A Title Deed (*Nor Sor 4 or Chanot*) indicates full ownership, and is supported by a deed plan showing the position of boundary mark stones. A Certificate of Utilization (*Nor Sor 3 or Nor Sor 3 Kor*) proves that the person named as owner has put the land to use. A Pre-emptive Certificate (*Bai-Chong or Nor Sor 2*) authorizes temporary occupation of land. A Claim Certificate (*Sor Kor*

1) indicates a claim lodged by a person who was in possession and had made use of the land prior to 1954.

After the Second World War, rapid expansion of cultivated areas continuously took place through the clearing of forest areas, without authorization and formalization of farmers' land rights. In the early days, squatters' rights over the forest land transformed into farmland were never recognized. When the problem reached critical proportions, the government adopted two main guidelines to solve the squatter problem. First, confronted with a *fait accompli,* encroachment by farmers has often been backed by formal legislation, and pre-existing land occupation was validated and turned into land ownership or land occupation rights. Second, many government agencies were created to support land allocation programs on public land. There are currently some 14 different agencies carrying out their land settlement programs based on different laws and accordingly issue many forms of land titles. For example, the Department of Public Welfare issues *Nor Kor* which allows the settlers (after a certain period of time) to apply for certificates of utilization and later title deeds. The Department of Cooperatives Promotion issues *Kor Sor Nor* which is similar to *Nor Kor*. Since 1982, the Royal Forest Department has been issuing *Sor Thor Kor* for forest encroachers. This document grants squatters the right to occupy land in national forest reserves provided that the recipients do not hold more than 20 rai of land and carry out only agro-forestry activities. The Agricultural Land Reform Office issues *Sor Por Kor 4-01* which is a land use permit similar to *Sor Thor Kor*[4].

A multiple system of land ownership has thus developed, which is reflected in multiple sources of land documents granting different rights on the land cultivated. As far as exclusivity and transferability rights are concerned, farmers in Thailand have obviously different tenure security statuses[5]. Usufructuary rights to farmland were acquired by initial clearing and

4 Since the land reform scheme was introduced twenty two years ago, about nine million rai of forest land have been given to 460 000 applicants.

5 The meaning of *tenure security* can be unclear and confusing. When it refers to the ability to cultivate a piece of land in peace and to collect the income drawn from that land (exclusivity), security is usually ensured under traditional systems. The difficulty with customary system of land tenure usually lies in its lack of freedom of alienability. If tenure security is defined in terms of transferability rights, i.e. the ability of an occupant to undertake land transfers, the presence of a legal framework creates security in relation to the nature of farmers' rights over the land.

maintained by cultivation. According to the 1954 Land Code, full and unrestricted legal possession of the land is warranted only by the *N.S.4* or *Chanot*. The *Nor Sor 3* and *Nor Sor 3 Kor*, however, are in practice, taken as evidence of legal ownership, and as such, used as collateral for loans. All other documents issued by various government departments do not grant complete transferability rights; farmers who possess any kind of legal utilization right, even temporary, hold secure exclusivity rights over the cultivated land; land can usually be passed on to descendants of the holder for agricultural purposes but cannot be sold.

Under current Thai legislation, only tenancy implies a low level of certainty regarding continuing access to the cultivated land. The security of a specific arrangement or contract depends, to a large extent, on the basic institutional environment. In village communities this environment consists mainly of traditional customs and moral principles. If a landowner imposes a contract inconsistent with village customs and principles, he would lose his legitimacy as a *patron*. The tenancy contract is thus more easily enforced if contractors are incited to conform to social norms such as reciprocal rights and duties that are clearly defined by tradition. Although no quantitative data on the duration of tenancy contracts were gathered during the survey, our observations revealed that most tenancy contracts were established between relatives and neighbors. In this case, it would seem that a strong motivation should exist to preserve and strengthen relationships that involve highly personalized transactions. Hence, we assume that tenancy contracts lead on to stable employment opportunities on a long-term basis.

Finally, the issuance of documents has always been delayed by the failure of the state agencies to manage the system, to keep records complete and up-to-date. The high number of institutions dealing with land allocation, with overlapping powers and jurisdictions, has been responsible for delays, lack of firmness in the resolution of land rights, and manifold opportunities for favoritism. Furthermore, a titling program involves cadastral surveys and demarcation of boundaries, a process that is usually costly. As a result, in 1954, only 12 million rai were registered, representing less than 35% of the cultivated area (Kemp 1981). Despite great improvements since 1954, a significant percentage of private land was still not properly documented in the early 1980s; only 12% of occupied agricultural land was covered by title deeds, 49% was held under less secure forms of land right document, 18% was occupied by persons who lack any officially recognized documentation, 21% was illegally forest reserve (Onchan, 1990; Rattanabirasongse, 1998). In 1984, Thailand engaged in a large land titling project which benefited over ten million people. What has been the impact of this land policy on rice farmers?

Table 1.2 shows that most rice farmers surveyed in the present study who claimed ownership land actually own plots of land for which they have a document proving full ownership. There is little private land that is not properly documented in the areas surveyed. The *Chanot* is by far the most dominant title owned by farmers of Suphan Buri and Roi Et. The *Chanot*, the *Nor Sor 3* and the *Nor Sor 3 Kor* are more equally distributed in Pitchit. Those who do not possess documents for the land they claim to own are mainly located in Roi Et, while this case is quite infrequent in Central Plain.

Table 1.2: Distribution of Land Titles by Province

	Suphan Buri	Pitchit	Roi Et	All Provinces
Chanot	65	20	61	146
Nor Sor 3 Kor	7	24	19	50
Nor Sor 3	5	33	7	45
Baichong	4	4		4
Sor Kor 1		2	1	3
No document	3	7	16	26

SOURCE: ORSTOM/CUSRI Project, Survey on Rice Farmers

Because rice farmers have documents giving them different rights to different pieces of land, landowners can be further be categorized in three distinct classes in terms of their tenure security status: the first group comprises fully secure titled farms which possess a secure title on each tract of the land they cultivate. Partly secure titled farms refer to those farmers who operate both legally owned land for which they possess a proper document and land for which they do not hold full transferability rights. Unsecured farms do not hold any document or titles that provide full ownership of the land they cultivate.

Table 1.3: Distribution of Ownership Security by Landownership Size and by Province

	<15 rai	16-30 rai	31-50 rai	51-100 rai	101+ rai	All sizes
Suphan Buri						
Fully secure titled farms	39.5	82.4	100.0	100.0	100.0	64.0
Partially secure titled farms	55.3	17.6	0.0	0.0	0.0	32.0
Unsecure titled farms	5.3	0.0	0.0	0.0	0.0	4.0
Pitchit						
Fully secure titled farms	55.3	65.4	63.3	76.9	83.3	63.8
Partially secure titled farms	33.3	11.5	31.8	7.7	16.7	20.5
Unsecure titled farms	13.3	23.1	4.5	15.4	0.0	15.7
Roi Et						
Fully secure titled farms	72.7	60.0	73.3	100.0	100.0	70.3
Partially secure titled farms	10.9	28.0	0.0	0.0	0.0	12.9
Unsecure titled farms	16.4	12.0	26.7	0.0	0.0	16.8

SOURCE: ORSTOM/CUSRI Project, Survey on Rice Farmers

An interesting breakdown of land titles by size of landownership in Table 1.3 shows a positive correlation. In Suphan Buri, only 40% of small owners possess documents providing complete transferability rights *versus* 100% of owners of 31 rai and more. Although this gap is lower in Pitchit and Roi Et, there is also a tendency for larger owners to combine land ownership and a high degree of security for the land they own. Similarly, Table 1.4 suggests that those who obtained titles tend to be older and slightly better educated (and usually well informed) than unsecured farmers[6]. In general, the household heads who own secured land have more frequently completed lower elementary school, while 30% to 40% of heads operating untitled land did not attend school at all or left before the end of the first four years.

These results strongly suggest that, apart from the implications for productivity, ownership security also raises questions of equity. The establishment of property rights as well as the procedures required to prove legal ownership are always complex and involve significant transaction costs. During the survey, farmers often complained about the delays and costs involved before they could get through the whole official procedure. Since these costs vary little according to the size of the farm, larger and wealthier landowners are better placed to afford them. Transaction costs have thus biased the acquisition of titles in favor of large and wealthy farmers. Furthermore, because of differences in educational levels and, consequently, differences in access to the state administration, some segments of the population are exposed to the risk of exclusion from access to land because they are unaware of the implications of registration or are unable to have their existing land rights recorded. The history of Thailand reveals that in many cases land records have been manipulated by powerful government officials to allow elite to obtain ownership of land (Feeny, 1982).

[6] Measuring educational levels however, is not very meaningful because very few farmers have any schooling higher than primary school.

Table 1.4: Characteristics of the Household Head and Security of Ownership

	Fully secure titled farms	Partly secure titled farms	Unsecure titled farms
SuphanBuri			
Head Age	56	52	48
Education level (% of farmers)			
No formal edication	25.0	8.3	33.3
Lower elementary	72.9	83.3	66.7
Upper elementary and more	2.1	8.3	0.0
Pitchit			
Head Age	53	48	49
Education level (% of farmers)			
No formal edication	26.4	33.3	38.5
Lower elementary	73.6	61.1	61.5
Upper elementary and more	0.0	5.6	0.0
Roi Et			
Head Age	51	49	48
Education level (% of farmers)			
No formal edication	21.7	23.1	0.0
Lower elementary	78.3	76.9	100.0
Upper elementary and more	0.0	0.0	0.0

SOURCE: ORSTOM/CUSRI Project, Survey on Rice Farmers

1.2 Access to Land Ownership: The Challenge

Generally speaking, access to land ownership is bound to be affected by a rise in population numbers. The combination of an increasing rural population and a naturally limited land base should lead to a reduction in the average size of owned holdings and to land fragmentation or scattering, which is the result of subdivision of shares by each new generation and by increasing numbers of co-successors. If the average rate of population growth is the same among large and small landowners, families with initially larger land endowments would be better off than the others; partitioning by the smallest holders would tend to produce holdings too small to provide a living, forcing them to find additional or alternative sources of employment. Even if land owned/person ratios were stabilized, the agrarian structure could still be pushed toward a unimodal distribution of large farms because small farms are unlikely to be competitive with larger farms due to scale economies in both production and market access. Whichever explanation is correct, disparities in the ownership of land should increase over time, due principally to downward mobility resulting from land sales by smaller owners to larger ones.

1.2.1 Concentration in Rice Land Ownership

Long-term changes in the overall degree of land concentration in Thailand are not easy to assess since there is a lack of empirical data prior to the first agricultural census in 1963. However, available empirical evidence tends to suggest that initial agrarian structure was characterized by a low incidence of landlessness. There is little indication that officials and royalty maintained vast estates consistent with their rank, and that the government land policy consisted chiefly of encouraging cultivation (Kemp, 1981). Although Thailand remained an agricultural country until the end of the Second World War, land, especially rural land, was of minor concern as a political resource to the traditional *élite*. The major sources of official revenue was the *corvée* labor required for all active freemen, goods and services supplied by their *clients* and later from trade.

Furthermore, it is questionable whether the inequality of land ownership significantly deteriorated over the decades. An interesting work on the Central Plain pointed out that rural conditions did not seriously deteriorate during the twentieth century and that the ownership pattern did not worsen between 1930 and 1970 (Stifel, 1976). Afterwards the data from the agricultural censuses (1963, 1978 and 1993) and from the intercensal surveys of agriculture (1983 and 1988) do not show any tendency for land ownership to concentrate over time. The Gini coefficients calculated from these sources were 0.46 in 1963 and in 1978, stabilized around 0.36 in 1983 and in 1988, and diminished to 0.30 in 1993. The concentration in land ownership has thus slightly decreased since the 1970s.

Almost all rice households in Thailand own at least some land. Unlike other developing countries, the incidence of landlessness is remarkably low.

However, between 1963 and 1993, the proportion of all rice farms owning no land at all increased from 3.8% to 10.4% as well as the share of medium-sized holdings (10-40 rai) from 44.2% to 56.9%. The share of all other groups remained almost unchanged. These results suggest that many owners have succeeded in avoiding the subdivision of their land while those who had previously owned the smallest pieces of land have lost them. The *surplus* of land generated by this loss was acquired by the medium-sized farms. However, there is no clear empirical evidence showing a dramatic increase in these workers completely dependent on the market for wage labor, usually called landless farmers. According to the Labor Force Surveys, only 11.6% of the agricultural labor force relied wholly on agricultural wage labor in 1999, against 10% in 1990. The absence of a landless group is partly explained by the development of land rental markets that have been an important means of redistribution of cultivated land. The share of total area under tenancy contracts grew from 4% in 1963 to 12% in 1993.

The data from the field survey gives a slightly different picture. Although there are very few large landowners among rice farmers, holdings of own land remain quite concentrated (Table 1.5). There are, however, large differences in the structure of land ownership between provinces. The proportion of households owning no land varies from a low 5% in Roi Et to a quite significant 26% in Suphan Buri. The smallest average area owned is 18.8 rai per owner household in Roi Et, and the largest 42.5 rai in Pitchit. The percentage of large own holdings is the highest in Pitchit: 20% of holdings own more than 50 rai compared to 7% in Suphan Buri and 3% in Roi Et. Conversely the proportion of holdings owning less than 15 rai is 53.4% in Roi Et compared to 38% in Suphan Buri and 15.3% in Pitchit.

In Suphan Buri where the distribution pattern of land ownership is the most skewed (the Gini coefficient is equal to 0.60), the 64% of farmers owning less than 15 rai account for less than 20% of the area owned. At the other end of the scale, 7% of households own more than 50 rai and account for nearly 40% of the area owned. In Pitchit and Roi Et, the ownership of land appear less concentrated as reflected in a lower value of the Gini index (respectively 0.2 and 0.40).

Table 1.5: Distribution of Land Ownership by Farm Size

	No land	<15 rai	16-30 rai	31-50 rai	51-100 rai	101+ rai	Average area owned
Suphan Buri							26.0
Percent of farms	26.0	38.0	17.0	12.0	6.0	1.0	
Cumulative percent	26.0	64.0	81.0	93.0	99.0	100.0	
Percent of area		17.9	20.0	25.0	26.5	10.4	
Cumulative percent		17.9	38.1	63.1	89.6	100.0	
Pitchit							42.5
Percent of farms	16.3	15.3	26.5	22.4	13.3	6.1	
Cumulative percent	16.3	31.6	58.1	80.9	93.9	100.0	
Percent of area		4.1	16.4	25.3	24.8	29.4	
Cumulative percent		4.1	20.5	45.8	70.6	100.0	
Roi Et							18.8
Percent of farms	4.9	53.4	24.3	14.6	1.9	1.0	
Cumulative percent	4.9	58.3	82.6	97.1	99.1	100.0	
Percent of area		26.0	29.2	31.3	6.4	7.0	
Cumulative percent		26.0	55.2	86.5	92.9	100.0	

SOURCE: ORSTOM/CUSRI Project, Survey on Rice Farmers

In summary, the evidence presented above shows that much land has remained in the ownership of small peasant farmers. Four main reasons may explain this remarkable achievement. First, for a very long period of time, labor was an asset relatively more valuable than land. Second, because of the large amounts of virgin land to clear, inter-village migration in search of cultivable and unoccupied land as a method of distributing excess population in areas with excess land was a recurrent feature of Thai peasant society. Third, all land laws enacted in Thailand since the beginning of the twentieth century, granting ownership to the actual operators, has resulted in a distribution of land ownership which is, by international standards, quite egalitarian. Fourth, farmers have been able to compensate for the lack of land due to population pressure by partitioning land unequally among their heirs. With the gradual end of the land frontier, there has been a tendency to give all the land to one child mainly because farmers are reluctant to divide their holdings below what they consider economically viable sizes (Mehl, 1986)[7]. Fifth, the country has and has enjoyed an impressive economic development for about the last four decades that resulted in outward migration of "surplus" labor.

1.2.2 Land Transactions

In theory, the land market is viewed as a mechanism that facilitates land circulation and fosters more efficient resource allocation in agriculture. A perfect land market should tend to allocate land from households with relatively low endowments of other factors to households with relatively high endowments of these factors. From the purchaser point of view, land acquisitions are made both to increase profits from cultivation and to leverage further land expansion through offering a suitable collateral to lending institutions. Hence, a high rate of return to land provides a strong incentive to accumulate more land, especially where capital markets are underdeveloped. Conversely, voluntary land sales will occur only when sellers are expecting a potentially higher return from the proceeds of the transaction than from self-cultivation or land rental. Thus the most important factors contributing to the

[7] This contradicts the traditional practice which has long been to give equal land rights to children, despite some regional variations and sometimes preference for the child who takes care of the parents (Sanit and Prisana, 1988).

development of a land market are the use of new technology and inputs increasing the returns per unit of land, the development of markets for products, and the rise of alternative uses for land.

Land transactions are long-term decisions and more than one purchase and/or sale may be made by a farmer over a lifetime. During the survey, farmers were asked to report about their land sales and purchases[8]. Responses are recorded in Tables 1.6 and 1.7: 37 households sold at least one piece of land and 96 purchased at least one over the period from 1937 to 1995. The land market did not appear very active, with on average less than one transaction per village and per year. On the whole, when they are transferred, the majority of parcels continue to follow the path of customary channels, among which inheritance is the most important. This tends to indicate that land purchase has not been a predominant means of personal farmland enlargement compared to inheritance.

Several reasons for the low turnover in the market for land must be considered. First, in agrarian economies exposed to various kinds of risks, land is an attractive asset to hold even if the rate of return on the investment happens to be low because of the very limited existence of other objects of investment. Land represents not merely a primary farm input, but also an easier access to credit, a preferred means for storing wealth, a speculative asset and income security. For many Thai peasants, land still represents the ultimate security and they are reluctant to give it up, a decision that would leave families and their descendants landless. A second factor likely to reduce the trade flow in the land market is the opportunity of seasonal non-agricultural employment for rural residents. Most Thai farmers are engaged in off-farm activities to a certain extent as a means of matching land and labor and of temporarily absorbing agricultural labor surplus, thereby reducing the need for land adjustment. Third, Thai farmers relied on alternative mechanisms, such as land renting (in and out), to match land and labor. A fourth explanation is the reinforcing nature of the low turnover: since the volume of transactions is a good indicator of one's chances of being able to

[8] Most households coped with only one land transaction during their life cycle. For the very few households which made multiple land transactions, only the most recent one has been recorded. Hence, the last land transaction reported is quite a good indicator of the market activity. In addition, land sales are certainly under-estimated. Those who sold all the land they once owned and joined the ranks of *landless farmers* or agricultural wage workers were by definition not questioned during the survey.

make a transaction, individuals hesitate to sell land when land turnover is low, because there is no guarantee of a seller being able to buy back land in the immediate future (Basu, 1986). Land sales are low precisely because land is the last asset people part with. If farmers were able to cash their land assets temporarily in some way, the benefit of a high turnover that these sales would induce would be similar to the benefit of having more credit in an economy. Fifth, high transaction costs of land exchange constitutes an important factor that hinders trade in the land market. It is lengthy and costly for a farmer to go through all the necessary transactions in order to have a large and consolidated plot. This discourages farmers from attempting to enlarge the size of the operation through land market transactions. Sixth, the farmer's purchase of additional land is restricted by both the savings he decides upon and the loans he can obtain, the latter depending on the collateral he can offer to secure loans. As rural Thai credit markets function imperfectly for a variety of reasons, including moral hazard and adverse selection, the number of land buyers is restricted by the lack of access to credit.

The frequency of land transactions varied from province to province. In Suphan Buri, 60% of the farms were engaged in the land market, a higher proportion than that found for the other two provinces. This greater occurrence in land transactions in Suphan Buri compared to the other provinces is explained by the greater integration of this area into the market economy, the greater accessibility to markets, greater urbanization and thus greater incentives to convert agricultural land into urban uses.

That households with larger financing capacities were better able to acquire (additional) land is confirmed by the distribution of land purchases among classes of land ownership, which highlights a predominance of big owners. In relative terms the largest owners reported land purchases more often than those owning small holdings. Also, there is a positive relationship between the land owned and the average area purchased, those owning more land having purchased the largest areas.

Farmers also exhibit a strong preference for using own savings to purchases their land: 78% of the land purchases reported were financed from own savings in Roi Et, 57% in Suphan Buri and 50% in Pitchit. Furthermore, farmers owning the largest landholdings turned to credit for their financing of land purchases more often than farmers owning small plots of land.

Table 1.6: Distribution of Land Purchases by Holding Size

	No land	<15 rai	16-30 rai	31-50 rai	51-100 rai	101+ rai	All Sizes
Suphan Buri							
Frequency	2	9	12	8	5	1	37
Percent of size class	7.8	23.7	70.6	66.7	83.3	100.0	37.0
Average area purchased	15.0	8.9	15.9	91.5	41.2	200.0	38.9
Pitchit							
Frequency	2	4	9	8	10	3	36
Percent of size class	12.5	26.7	34.6	36.4	76.9	50.0	36.7
Average area purchased	25.0	8.8	11.8	35.1	50.6	90.0	34.7
Roi Et							
Frequency		6	8	6	2	1	23
Percent of size class		10.9	32.0	40.0	100.0	100.0	22.3
Average area purchased		3.5	11.6	15.5	23.0	13.0	11.6

SOURCE: ORSTOM/CUSRI Project, Survey on Rice Farmers

The main reason for land purchases to be financed from savings is related to the use of land as collateral for institutional and non-institutional loans as well (Binswanger and Rosenzweig, 1986). This argument goes in two directions. On the supply side, where lenders insist on collateral, the credit market is not open to those who do not own assets acceptable as collateral, or is open at a prohibitive interest rate. On the demand side, small farmers reported that they had limited access to any kind of credit because of their weak assets base and, consequently, were forced to self-finance their land purchases. In addition, if the land purchased with a loan is used as collateral for that loan, the land cannot be used again as collateral for working capital. The borrower might not be able to secure and increase his income stream produced from the land and might fail to repay the loan. Thus purchasing land by obtaining a loan would be unprofitable.

The frequency of sales varies from province to province. The supply of land for sale is lower where non-agricultural investment opportunities for rural residents are limited than in areas where the possibility of alternative uses increased the value of land. Consequently, the occurrence of land sales is the highest in Suphan Buri. In relative terms, the frequency of land sales is much higher for big owners than for small ones in Suphan Buri and in Pitchit, while in Roi Et all land sales are made by farmers owning less than 15 rai. Furthermore, the average area sold tend to increase with farm size, indicating that land sales are not likely to exacerbate landholding inequality.

One question that often arises is the reason for the sale of the land. Land may change hands in three main ways: through foreclosure in case of loan default, through distress sales or through voluntary market transactions. The most frequent reason for land sales reported by farmers were primarily loan repayment (40% of answers) and secondly family needs (30%). This second reason rests critically on the fact that cheap institutional credit is not usually available for compensating consumption or temporary income falls. Thus, when a household is faced with an unexpected shock, it has to sell off some of its assets, for example land. If cheap institutional credit were available, then households in need of cash could borrow and postpone or cancel the sale of a plot.

Table 1.7: Distribution of Land Sales by Holding Size

	No land	<15 rai	16-30 rai	31-50 rai	51-100 rai	101+ rai	All Sizes
Suphan Buri							
Frequency	5	9	5	1	3	0	23
Percent of size class	19.2	23.7	29.4	8.3	50.0		23.0
Average area sold	14.2	12.4	26.4	18.0	10.7		15.9
Pitchit							
Frequency	1	1	3	2	2	2	11
Percent of size class	6.3	6.7	11.5	9.1	15.4	33.3	11.2
Average area sold	15.0	30.0	13.1	25.5	15.0	44.0	23.0
Roi Et							
Frequency	2	1					3
Percent of size class	40.0	1.8					2.9
Average area sold	0.3	5.0					1.8

SOURCE: ORSTOM/CUSRI Project, Survey on Rice Farmers

Other reasons (30%) for land sale are associated with the practice of equal inheritance which tends to subdivide land area eventually to the point of uneconomic operation, resulting in the sale of land. Bad land location is another reason which drives many farm households to seek to consolidate their holdings by occasionally exchanging plots with one another. Finally, comes the critical and definitely non-traditional decision to sell land for investing in more and better quality education for the children instead of giving them land or a substantial bride price which would allow them to purchase land. This may be considered as a portfolio re-allocation decision which shift assets from land to human capital.

An overwhelming proportion of land transactions occurred among farmers, whatever the province. Opposite to a widespread opinion, very few areas have in fact been sold to merchants or private moneylenders. Around 40% of those who exchanged land were transferring it to other family members (close or distant); the other 30% of land transactions occurred between neighbors and another third between unknown people.

Table 1.8: Land Transactions and Operators Characteristics

	Stephan Buri	Pitchit	Roi Et
Merchants	4	4	
Farmers	50	34	23
Others	7	4	
Relative	24	10	18
Neighbor	15	24	1
No Relation	22	8	4
Total	61	42	23

SOURCE: ORSTOM/CUSRI Project, Survey on Rice Farmers

These empirical findings suggest that a farmer's acquisition of additional land is often associated with the demand for land consolidation consistent with the pattern of land inheritance. When the plot inherited is too small to be efficient, it is not uncommon for one of the children to buy their fragments of the farm from the other siblings (Mehl, 1986). The predominance of family transactions may be also the manifestation of the gains obtained through elderly kin-based knowledge (Rosenzweig and Wolpin, 1985). The accumulation of useful information is most valuable when applied continuously on particular parcels of land. The experienced elderly can supply information about the most efficient techniques for coping with previously experienced various adverse conditions, techniques that are likely to have a farm-specific component. The existence of returns for specific experience thus creates incentives for farm descendants to purchase that land. However, the *experience effect* is most likely to occur in an environment with stagnant technology. Accordingly, the non-family land market is more severely restricted in Roi Et where kinship ties are strongest and technology stagnant than in the more advanced urban peripheries of Suphan Buri.

Similarly, land sales among non-resident operators limited because land is not homogeneous, and knowledge about the quality of a particular plot cannot be easily obtained. Buying land requires an ability to acquire and use new information. Transaction costs associated with information costs are likely to discourage land sales to people outside the village. Since proximity partly solves this problem, potential buyers are predominantly living in the village of residence of the seller.

1.2.3 Land Prices: An Absolute Constraint to Land Property

At the heart of the literature on land values is the asset pricing theory. The simplest version of this theory assumes land buyers value land only for its economic return. If this assumption holds, then a close relation exists between land prices and land rents. The price of land should thus be positively related to its productive potential over a long run, i.e., on expected returns or rents. However, the price of land as conceptually treated in the Ricardian rent theory is the price of *pure* land, independent of environmental characteristics and local public goods embodied in land. In fact, the value of agricultural land is obviously related to a wide range of factors: productive quality, availability of water, access to infrastructure and distance from market, title status, and so on. The capitalization of all these elements may drive the price of farmland far above the present value of expected agricultural profits.

Attention has recently been paid in the literature to the value of environmental quality, suggesting that land values vary with accessibility. Land is composed not only of characteristics relating to the productive quality itself, but also of characteristics determined by location. The latter includes a wide range of location-determined characteristics such as the general condition of the local economy, the quality of the micro environment, the distance to markets or to main roads, and local public goods. These characteristics are not specific to any particular land parcel but rather to all land in a particular area and are expected to be capitalized into the value of land.

The price of farmland may also depend on the debt it can carry. Because assets with collateral values are often necessary for access to credit, one might expect farmers with relevant land documents (*Chanot, Nor Sor 3, Nor Sor 3 Kor*) to get a higher price than farmers owning untitled land of identical quality. The price of titled land should include a premium that reflects both the expectations of capital gains and the usefulness of land as a form of collateral.

During the survey, farmers were asked to assess the market value of the land they own. Table 1.9 records the mean prices of land reported by rice farmers[9]. Since the size of the data pool allowed for the neutralization of region specific influences, the data were broken down by district, by registration status and by a proxy of the productive potential of the land, namely irrigation[10]. The results support several commonly-held hypotheses about land prices. Land markets in different areas of Thailand, and accordingly land rents and land prices, appear to vary greatly. Land prices are indeed prohibitively expensive in Suphan Buri with an average of 60,000 baht per rai, while the estimated value given by farmers for their own land ranges from 25,000 baht in Pitchit to 10,500 in Roi Et (1995 prices). Accordingly, land rentals vary from 560 baht per rai in Suphan Buri, to 344 rai in Pitchit and 254 baht in Roi Et (cf. infra). Land values thus tend to have the same cross-sectional representation as land rents. However, the premium in Suphan Buri's land appears far larger than the premium observed in land rents. Land values estimated by farmers widen the existing gap between current returns.

[9] It is implicitly assumed that land owners were accurate in assessing the value of the land they own.

[10] The productive quality of land is difficult to assess both theoretically and empirically since there is a wide range of criteria which can define it. Lack of appropriate data in the field survey concerning the agronomic quality of land prevented any more detailed investigation.

Table 1.9: Average Estimated Land Values

	Stephan Buri		Pitchit		Roi Et	
	Yan Yao	Kraseaw	Tayeam	Taluang	Mava	Pajao
Rice only	56750	67500	22429	23550	9963	10233
Rice and other cereals	51666		25824	19500		
Rice and cash crops	62250	52777		20000	9333	10000
Rice and fruits	62777	53666	31357	38600		
Rice and livestock and fisheries	53666	58750		31500	13917	11563
Dam irrigation	55000	61400	27611	26270		
Dam and pumping irrigation	67000	56875	25000	42000		
Pumping irrigation			23750		12500	
No irrigation			26400	18000	10456	10426
Fully titled farms	62000	55809	30750	27921	11037	10847
Partially titled farms	54286	62500	29214	16166	9333	9692
Untitled farms	35000		21167	14333	9083	9417
Total average	58142	57655	25197	25207	10569	10418

SOURCE: ORSTOM/CUSRI Project, Survey on Rice Farmers

Treated as site characteristics, i.e. factors that are more or less common to all land parcels in a given village, access to irrigation facilities have a powerful influence on the local average price of land. Between provinces, the access to better irrigation systems is associated with higher values of land. Within each district, however, the effect of irrigation systems on land values is not easy to determine; this is particularly true for the land irrigated by both dam and pumping compared to land irrigated by dam. The main reason for this result comes from the fact that land irrigated by pumping is usually partly devoted to crops other than rice, a factor responsible for higher prices of land. In any case, land which is not irrigated at all is always much less expensive than land which is irrigated or watered.

As a strong link between land values and the land use pattern is expected, the effect of suitability of land for other crop cultivation on the price of land is of particular interest. Anthropologists argue that farmers give their paddy land higher values than land devoted to other crops because of the central role of rice in Thai cultural and social life (Mehl, 1986). This does not seem to hold true. Within each province, rice land was generally estimated to be worth less than a combination of rice land and upland. This reflects the higher economic benefits that are generated by crops like sugarcane, cassava, maize, and fruit trees. Even in Suphan Buri, year-round irrigation allowing double cropping and avoidance of the risk of a water shortage does not bring about higher values of paddy land compared to a combination of rice land and upland.

Land further from the urban fringe has a lower value because it is too far away from economic activity to be considered for its value in future urban uses. Since land in Suphan Buri is the closest to the more developed cities, the price of any particular parcel has been influenced by proximity to infrastructure and to commercial and residential development. In this province, the possibility of conversion to nonfarm use increased land values. This is reflected in a higher average price of land. Conversely, land in Roi Et is at too great distance from Thailand's main urban centers to have any value in potential urban uses, and the speculative value of land is low.

Finally, as farm productivity is expected to be related to land ownership security, titled land must have a higher price than untitled or unsecured titled land of identical quality. It is perfectly clear that holding a legal title is an important factor in explaining differences in land values. Land with a secure title is always more expensive than land with an insecure title, which in turn, is substantially higher than the value of land without any document proving ownership of land.

Variations in Thailand farmland prices attracted widespread interest with their recent alleged explosive appreciation. Many explanations have been advanced in the literature to explain farmland price movements. Any shift of farmland price can emerge from demand or supply shifts, or from some combination thereof. The demand for agricultural land usually shifts up because of an increase either in farm product prices or in the productivity of land. A major reason for supply shifts is increases in nonagricultural demand for land, particularly for urban use, which drives land out of farming. Although our empirical knowledge is yet far from sufficient to construct a meaningful model explaining demand and supply characteristics of riceland in Thailand, we have attempted to present an analysis of the trend in farmland prices over the past three decades in the areas surveyed. These factors include changes in returns on farming, nonfarm returns on land and explosive expectations.

Land transaction prices were recorded during the survey, as well as the amount of land bought or sold. The set of all land transactions was reduced into a subset to include only those parcels that were at least 5 rai in size, and which were non-family transactions. Since individual parcels of land were sold at irregular intervals, the time-series used here refers to the average price of land at various points in time. In order to remove the influence of changes in general prices, land prices were divided by the consumer price index. The results are summarized in table 1.10.

Deflated riceland values declined from the 1960s until the beginning of the 1980s. Thereafter, a gradual increase occurred through the 1980s, followed by a marked acceleration in the beginning of the 1990s. Although farmland price trend was found to be rather identical when land prices are classified by nature of transaction, real selling prices have always been above real purchasing prices. In addition, the gap between these prices rose sharply in the late 1980s in Suphan Buri and Pitchit. Real selling prices showed a dramatic upward trend, while the real purchasing price increased much less, and even decreased in Pitchit.

Table 1.10: Average Real Land Transaction Prices (baht per rai)

	Stephan Buri		Pitchit		Roi Et	
	Real purchasing price	Real selling price	Real purchasing price	Real selling price	Real purchasing price	Real selling price
1964	564					
1965						
1966	4955		2126	4204		
1967			2042		744	
1968						
1969						
1970						
1971		3300	2063			
1972						
1973			2725		2725	
1974					3289	
1975						
1976					2493	
1977						
1978						
1979			1959			
1980			3929			
1981	3301	4358	2615	6963	1859	
1982					1795	

1983						
1984		3958			3729	
1985	5152		1546		3993	
1986		5227	1315	1290	4123	12645
1987		6417	1974	2592		
1988	7605	9506	2376		4753	7129
1989		12816	2708			
1990	17242	22988	745	18093	4257	10217
1991		28192		27381		

SOURCE: ORSTOM/CUSRI Project, Survey on Rice Farmers

The value of land as determined by the Ricardian theory depends first on the expectations land buyers have about future returns on rice farming activity. As a means of production, the demand for rice land should decrease when the rate of return on agricultural investment declines compared to other investments. No time series on real returns on rice farming are available. However, given the average low level of expenditures on intermediate inputs in most rice producing areas one might expect returns on rice production to have roughly moved in the same way as output prices. We can therefore consider that over the 40-year period from the 1960s up to now the rate of return on riceland declined due to constant decreases in real rice prices. While this is consistent with the trend in land values until the 1980s, land values do not seem well correlated with expected returns on rice farming in subsequent years, as plausible expectation schemes would dictate. From the 1980s onward, the major trend in real returns and real land prices are at opposites direction in all provinces.

The strong appreciation of land prices in the 1980s must be explained by factors other than rents that reflect the critical role of nonfarm uses of land. Although the area surveyed is primarily rural, most villages are surrounded by a number of large metropolitan areas, especially in Central Plain. Thus, multiple sources of urban influence have been encountered in the 1980s due to accelerated growth of the non agricultural sector resulting in the continued expansion of urban areas through industrial activity, residential development, second homes, recreational enterprises such as golf courses, which caused large amounts of farmland to be converted to urban use. The encroachment of urban areas has increased competition for land, resulting in rising land prices in the rural-urban fringe. However, the growth of rural industry has been uneven in the different regions. Rural industrialization has been highly concentrated in suburban areas near Bangkok, but has lagged far behind in remote regions. As a result, the tendency of land prices to overreact to changes in farming returns has been far more visible in Suphan Buri and Pitchit than in Roi Et.

The differences found between the buying and the selling land price reflect the fact that land prices differ according to expected use of land. It seems likely that land buyers were mainly farmers buying riceland for agricultural use, whose price is mainly influenced by expected farm income, whereas land sellers have sold parcels to persons wishing to acquire farmland for nonagricultural uses. Although this is an assumption that has yet to be tested, there is a strong presumption that the more favorably located land (with respect to the market) has been sold out of the agricultural sector while the less favorably located land remained inside the sector. Since in local markets, among land plots with the same characteristics, the more valuable land is the one located in the better position with respect to the market or to urban areas, selling prices must be higher than buying prices. The sale of land at market prices above those determined by the agricultural use value alone further increased the value of all land since prices convey information. Owners tend therefore to raise their expectations.

Aside from expected returns on farming and opportunity cost in nonagricultural use of land, land price expectations might have been an important explanatory force in the 1990s take-off period. Buyer expectations with respect to future development in the area where the land is located might have had a significant influence on land prices. If people expect land prices to rise, then it is logical for people to pay a price that reflects this increase. If land prices begin to rise at a faster than expected rate, then people will buy land, thus bidding up the current price. It might be possible that following the take-off period, part of the ensuing land price appreciation up to now was due to a positive effect on price expectations. These higher price expectations, in turn, caused a higher price the following year and so on.

In summary, farmers encounter great difficulties in buying land in Suphan Buri, because land prices tend to be much higher than the discounted net income they could earn from rice farming. This in turn is likely to slow down the growth of land market activity, reinforcing the low incidence of land transactions in the constitution of landownership. Conversely, farmers who do buy farmland in Pitchit and Roi Et appear less vulnerable to the divergence between land values and land returns.

1.3 Land Tenure: The Role of Land Rental Markets

Ownership size certainly carries significant implications for the level of expected income from rice farming, and as such the factors affecting the distribution of land ownership are important. However, looking beyond the distribution of ownership holdings helps to determine who has access to land and therefore who can benefit from strategies to make land more productive.

The leasing of land among rural households can be an important source of growing equality in operational holdings.

Tenancy contracts are quite widespread in developing economies. It is not easy to find an analysis of the economic factors that may explain variations in the incidence of tenancy in different areas. The causes of their prevalence and permanence are dependent on conditions under which tenancy is more desirable for the landlord than self-cultivation, while lease-in agricultural land is preferable for a landless farmer to working as a wage laborer. We have thus considered the forces involved in the market for land leases which affect the extent to which tenancy occurs.

1.3.1 Concentration in Rice Landholding Sizes and Land Tenure

Changes in land tenure in Thailand have run contrary to the general findings on developing countries. In many countries in the developing world, population pressure and the associated process of land concentration tend to increase the proportion of tenant farmers compared to owner farmers. In most cases, there was simply not enough land for the growing number of families in densely populated areas and landlessness developed as a "natural" consequence.

Three trends clearly stand out in the Thai national censuses. First, between 1963 and 1993, the proportion of households operating less than 5 rais declined slightly from 18.6% to 14.4% while their share in total area operated remained constant. Larger holdings increased proportionately in terms of both number and area. Second, inequality, as measured by Gini coefficients, has decreased somewhat over time; therefore tenancy has had an equalizing effect on the distribution of operational landholding. Third, in 1978[11], 85% of agricultural land was farmed by full owners, 5% was rented, the remaining 10% was operated under two tenure modes. In 1993 these proportion remained unchanged. The steady population growth in Thailand did not result in an increase of the landed as opposed to the landless classes. Nor did it induce any trend toward unequal private ownership of land or absentee ownership. This is because there were frontier areas in Thailand to which farmers without land

[11] The comparison from the census of 1963 would have been misleading since farmers holding land on two tenure mode were not isolated.

could migrate, and recent industrial development has created jobs to absorb the excess population.

The cumulative size distribution of operational holdings in the provinces surveyed is reported in Table 1.11. The results show that the rice sector is characterized by small farm family units and the proportion of households cultivating small holdings is relatively high, especially in the Northeast. Nearly 60% of the holdings in Roi Et are operating less than 15 rai while this proportion stands at 32% in Suphan Buri and 19% in Pitchit. This small farm structure, characteristic of South East Asian countries, is well adapted to the rice production which is very labor intensive.

The distribution of land operated among rice farmers is quite unequal. The distribution of operational holdings appears more skewed in Roi Et than in Suphan Buri and Pitchit, a fact confirmed by the decreasing value of the Gini coefficients from respectively 0.54 to 0.33 and 0.27. There are very few large holdings of more than 50 rai in the three provinces surveyed: only 1% to 4% in Suphan Buri and Roi Et, but 25% in Pitchit. Regionwise, the Northern region, with an average of about 42 rai, has proportionately larger holdings than the Central Plain (26 rai) and the Northeast (18 rai).

It must be remembered, however, that farm size is a relative concept, and its measurement in absolute terms can be quite misleading. Some large holdings can be very unproductive and operated at loss by the farmer while smaller holdings can be relatively more productive and operated at profit. Essential considerations are economic value of crop produced, soil quality, water management, and technology of crop production. A given farm size does not mean the same thing in an arid region like the Northeast as it does in an irrigated area like

Suphan Buri. Taking into account differences in the productive potential of land accentuates inequalities in landholding between provinces; the Northeast tends to combine both disadvantages of small and unirrigated holdings. Within the irrigated areas of Suphan Buri and Pitchit, the productive potential of farms is actually less unequally distributed than the sizes of the land operated, because there is a tendency for irrigated land to be concentrated in the smaller holdings.

Table 1.11: Distribution of Landholding Sizes

	<15 rai	16-30 rai	31-50 rai	51-100 rai	101+ rai	Mean Size
Suphan Buri						25.8
Percent of farms	32.0	42.0	22.0	3.0	1.0	
Cumulative percent	32.0	74.0	96.0	99.0	100.0	
Percent of area	12.4	38.6	32.9	8.3	7.8	
Cumulative percent	12.4	51.0	84.0	92.3	100.0	
Pitchit						41.8
Percent of farms	19.4	25.5	29.6	20.4	5.1	
Cumulative percent	19.4	44.9	74.5	94.9	100.0	
Percent of area	5.1	13.3	28.7	31.9	21.0	
Cumulative percent	5.1	18.4	47.1	79.0	100.0	
Roi Et						18.3
Percent of farms	58.3	21.4	19.4		1.0	
Cumulative percent	58.3	79.7	99.1	100.0	100.0	
Percent of area	27.9	26.4	40.1		5.6	
Cumulative percent	27.9	54.3	94.4	94.4	100.0	

SOURCE: ORSTOM/CUSRI Project, Survey on Rice Farmers

Based on the assumption that larger owners generally rent out land to smaller ones, it is usually expected that the size distribution of operational holdings is less skewed than that of ownership holdings. This presumed flow of land arise from the need to use land intensively, together with the diseconomies of scale associated with the management of labor. Comparison of figures in tables 1.5 and 1.11 suggests that the average area operated by the rice farmers sampled is not significantly different than the average area owned. Second, when comparing the distribution of land owned and cultivated, it appears that the concentration of landownership is much higher than the concentration of the landholding, as indicated by the Gini coefficients, which are respectively valued at 0.61 and 0.33 in Suphan Buri, 0.52 and 0.27 in Pitchit. In contrast, renting land does not compensate much for differences in landownership in Roi Et, but rather exacerbates them. These results can be interpreted as an indication of the positive functions of tenancy in Suphan Buri and Pitchit: providing non-landowners with access to land, and small and medium-sized farmers with the means to expand their farm size. In the Roi Et case, it seems likely that owners of small holdings, finding them uneconomic to run, rent them out to larger operators, and choose to earn their livelihood in some way other than farming. Consequently, many of those with relatively large operational holdings are tenants of small owners (cf. infra).

Rice farmers have been differentiated in table 1.12 into four broad tenure classes, according to the amount of land owned and rented. Landlord-owner farmers refer to those farmers who rent out part of their holding and cultivate the rest themselves, sometimes with rented-in land[12]. Owners are farmers owning all the land they cultivate, though documentary evidence of property may not yet be forthcoming. Owner-tenants own at least a portion of the farm land base while tenants rent in the total amount.

The first clear result is that owners and tenants cannot be identified as mutually exclusive social classes. Rather, villagers appear tied to one another through permanent market relations for land (and labor), resulting in a continuous spectrum of classes. There are however marked differences in the proportion of farmers by land tenure status between provinces. Owner farmers are the largest category in both number and area. They predominate in Roi Et where more than 70% of households own all the land they farm. The farmers

[12] In our sample 22 farmers fall into this category, among which two are renting in nearly the same area as they are renting out. The main reason lies under the fact that the land they own is too far away from home.

who own and cultivate land themselves account for 58% of the total number of farms in Pitchit, whereas this percentage drop to 37% in Suphan Buri. The ratio of ownership is thus higher in rainfed upland areas than in irrigated lowlands. Very few farmers (less than 10% whatever the province) reported renting out part of their holdings, and contrary to popular belief, households which rented out some land mostly belonged to the lower/medium holding size groups. These households do not appear to be leasing out their land because they have above-average holdings of land, but rather because they are pursuing higher income-earning opportunities in non-farm activities (cf. Chapter 2).

The figures in Table 1.12 also points to the existence of many holdings comprising more than one form of tenure. Owner-tenants make up the second largest group after pure owners. On average this group tend to operate the widest areas, supporting the argument that tenancy functions as a means of expanding farm size. About a quarter of farmers in Suphan Buri and Pitchit have expanded their farm land base by renting additional land, most of them owning as much land as they rent.

Finally, the level of tenancy is quite low if we consider farmers operating only land they do not own. The proportion of pure tenants ranges from about 4% in Roi Et to 14.3% in Pitchit and 25% in Suphan Buri. In terms of land area, the rate of tenancy, i.e. the proportion of rented land to the total area operated, appears to be far more pervasive around Suphan Buri (43.4%), compared to 25% in Pitchit and 17% in Roi Et. There is a danger in identifying tenancy as an indicator of rural poverty. Tenancy is not always associated with small holdings: in Suphan Buri, where tenancy has increased, the highest frequency of full tenants is in the 16-30 rai size class. Except in Pitchit, the average size of operational holdings are not typically much larger for owners than for tenants. In each province, the proportion of part owners/part renters tend to increase from small size classes to large size classes.

Table 1.12: Land Tenure by Landholding Size

	<15 rai	16-30 rai	31-50 rai	51-100 rai	101+ rai	All Sizes	Average area (rai)	Tenancy Rate (%)
Suphan Buri								43.4
Landlord-owners	9.4	11.9	22.7	-	-	13.0	26.1	
Owners	53.1	26.2	27.3	66.7	100.0	37.0	26.8	
Owners-tenants	15.6	28.6	31.8	33.3	-	25.0	27.6	
Tenants	21.9	33.3	18.2	-	-	25.0	22.4	
Pitchit								25.2
Landlord-owners	5.3	-	6.9	10.0	-	5.1	43.4	
Owners	63.2	64.0	51.7	55.0	60.4	58.2	40.8	
Owners-tenants	-	12.0	34.5	35.0	40.0	22.4	58.4	
Tenants	31.6	24.0	6.9	-	-	14.3	19.7	
Roi Et								17.3
Landlord-owners	6.7	4.5	20.0		-	8.7	22.1	
Owners	80.0	77.3	50.0		-	72.8	15.6	
Owners-tenants	8.3	18.2	25.0		100.0	14.6	29.7	
Tenants	5.0.	-	5.0		-	3.9	16.5	

SOURCE: ORSTOM/CUSRI Project, Survey on Rice Farmers

1.3.2 Factors Affecting Tenancy

The pattern and the incidence of tenancy is expected to be influenced by the structure and the nature of input markets. In theory, if markets for inputs were functioning perfectly, there would be no need for land tenancy. Under perfect market conditions, sales and purchases would move land to the complementary factors. We have seen that land transactions in the area surveyed are uncommon for many different reasons. As a result, most households are unable to adjust their land endowments to their optimal levels by selling or buying land. Therefore the distribution of operational farm sizes has been achieved through the land lease market where households having surplus resources in relation to their land lease-in land, while households with excess land in relation to their factor endowments lease-out land. In such a case, tenancy promote efficiency in second-best environments, i.e. the efficiency with which resources are allocated when one or more markets are functioning imperfectly.

We use a body of data in Table 1.13 to show how households adjusted the size of their cultivated land area through land rental market according to their resources in land, in fixed capital such as bullocks or tractors, and in family labor.

Let us begin with a preliminary remark regarding the quality of the land exchanged in the land lease market. The differences observed in the percentage of irrigated land cultivated between classes of farmers suggest that the land offered in the lease market is typically non-irrigated and hence of low quality. In Pitchit, for instance, where irrigation is evenly distributed, the irrigated proportion of cultivated area is the highest for households leasing-out land and the lowest for households leasing-in land. This means that farmers owning large areas in relation to their productive resources are more likely to cultivate the irrigated pieces of land because irrigation decreases the risk of cultivation and thus, indirectly, increases the returns from self-cultivation. The excess amount, which is likely to be non-irrigated, is offered in the tenancy market.

Table 1.13: Land Tenure and Households' Resources Endowment

	Landlord/owners	Owners	Owner/tenants	Tenants
Suphan Buri				
Percent of cultivated land which is irrigated	97.0	95.0	98.0	98.0
Land owned per worker	20.0	10.4	4.1	-
Land cultivated per worker	10.0	10.5	9.1	8.1
Age of household	57	56	52	53
Sex Ratio	1.08	0.97	1.02	1.33
Dependency Ratio	0.70	0.83	0.58	0.64
Total number of days worked off-farm	104	100	115	122
Value of farm machinery owned per worker	16251	5520	3141	4260
Pitchit				
Percent of cultivated land which is irrigated	80.0	70.0	62.0	57.0
Land owned per worker	33.9	14.3	12.9	-
Land cultivated per worker	14.5	14.3	19.5	9.2
Age of household	61	52	48	47
Sex Ratio	1.70	1.16	1.09	1.19
Dependency Ratio	0.38	0.82	0.7	0.65
Total number of days worked off-farm	60	70	50	83
Value of farm machinery owned per worker	7743	8214	12269	8165

Roi Et				
Percent of cultivated land which is irrigated	-	-	-	-
Land owned per worker	17.9	7.1	4.1	-
Land cultivated per worker	7.4	7.1	8.7	7.3
Age of household	60	50	49	42
Sex Ratio	0.70	0.99	1.22	0.75
Dependency Ratio	0.83	0.88	0.46	1.25
Total number of days worked off-farm	0	97	104	13
Value of farm machinery owned per worker	10928	1982	5279	45

SOURCE: ORSTOM/CUSRI Project, Survey on Rice Farmers

The differences in tenure between households is first explained by differences in labor endowments and labor characteristics. Labor availability for farming depends on the number of household members living on the farm, adjusted for sex and age. The composition of households becomes particularly important when social customs, legislation and/or market forces prevent some members from entering the labor market[13]. If there are no employment opportunities for some members of a household, except working on owned or leased land, this household can profitably lease in land to ensure fuller utilization of its family labor. In a longer run perspective, renting land will help the adjustment of land to labor for a given household as a result of changing family demographics. Over time, the family's demand for land first expand as children grow old enough to assist in the farm work and further contract in the latter phase of the family cycle as both obligations and labor force decline.

Accordingly, the characteristics of the land lessor should include advanced age, a small family labor force relative to the land owned, lower sex and dependency ratios, whereas the lessee should exhibit younger age, a larger family labor force relative to the land owned, and higher sex and dependency ratios. Table 1.13 indicates that the groups of pure tenants and of owner-tenants significantly increased the area per worker they cultivate compared to the land they own per worker through land rental markets. Conversely, landlord-owners decreased the land cultivated per worker compared to land owned per worker so that the average land cultivated per worker tend to be equal between groups of farmers. The observed average age of land lessors (around sixty) compared to that of tenants (from forty-two in Roi Et to fifty-three in Suphan Buri) is consistent with the family life cycle hypothesis. This age gap, nearly equal to one generation in Roi Et, suggests that younger households start with tenant farming, then acquire more land either through inheritance or purchases, and then become farmer-land lessor at an advanced age, reducing their farmed area by renting out land as the capacity of family labor decreases. Finally, our analysis failed to reveal any uniform influence of the households' composition on the amount of land leased in or out.

Third, desired landholding may differ between households because off-farm opportunities differ. Access to off-farm labor opportunities could result

[13] For example, children below a certain age cannot offer themselves for wage employment. Women also tend to spend less of their time working off-farm while men do so in a much greater

in lower desired landholding if off-farm activities are more remunerative than agricultural work. In such a case households will tend to allocate more labor to these activities and will have less labor available for rice cultivation; as a result the tenant's potential demand for leasing-in land will be lower. Table 1.13 shows that the number of days spent on off-farm activities is the highest in the group of tenants whereas the average area cultivated per worker is the lowest. There has thus been a tendency for tenant households to adjust the land cultivated according to the time devoted to farm and off-farm activities.

Fourth, the optimal size of operational holdings can also depends upon other factors endowments than labor, especially farm machines. We have included the value of farm machinery per worker as an explanatory variable of tenure. A greater availability of farm machinery is an indication of the household's ability to cultivate larger amounts of land and should induce cultivators to lease in more land. Table 1.13 reveals that households leasing out land have the highest investment in farm machinery per worker. This surprising result suggests that landlords employ a double strategy to save labor costs : leasing land out and investing in labor-saving equipment. The value of farm machinery is consistent with the size of land cultivated per worker between the three other groups of farmers. Owner-tenant households who cultivate the largest areas per worker also possess agricultural equipment whose value is above those of owner and tenant households.

Finally, general labor market conditions may induce (or restrain) landed farmers to enter land rental markets. Households who have excess land compared with family labor (landlord-owners) may find it more profitable to rent out land than to manage a large operation because of the scale of diseconomies arising from the use of hired workers and/or because the local labor market is under pressure. The landlord has the problem of motivating hired workers to work hard enough, and of supervising workers, two costs that pure family farms do not face. Because managerial ability is usually not tradeable, that is there are no farm managers who sell their services to other cultivators, landed households may find it preferable to lease out land.

Tenancy would also be high if the real wage rate is high, because it might be more profitable for the landowner to lease his land to a tenant rather than hiring workers. Looking at the relationship between the agricultural wage rate and the frequency of tenancy, it appears that the higher the wage rate for hired

proportion (cf. Chapter 2).

labor the more frequent the tenancy; in Suphan Buri a tenancy rate of 43.4% is associated with an average agricultural wage of 92 baht per day, whereas these figures are respectively 25.2% and 51 baht per day in Pitchit and 17.3% and 45 baht per day in Roi Et.

1.3.3 Landowner-Tenant Relationships

In theory, tenancy is considered as a second-best solution since it involves a principal-agent relationship and therefore gives rise to incentive problems such as asset mismanagement or labor shirking. However, tenancy can be efficient in resource use if the landlord can enforce the efficient level of work effort, which requires specific conditions. Generally speaking, landlords are expected to conclude contracts with tenants who are less risky and who require fewer enforcement costs. As a result, most tenancy contracts should be embedded in highly personalized relationships.

We therefore need to look into the nature of landlord-tenant relations in the study villages since they exert significant influence on the actual enforcement of contracts and thus the determination of tenancy form and conditions. The breakdown of tenancy contracts according to owners and tenants characteristics in Table 1.14 shows first that, on the whole, a majority of contracts were issued between farmers themselves. However, in Suphan Buri and to a less extent in Pitchit, a large number of landlords are part of the non-farming population. In these latter two provinces 52% and 36% of tenancy contracts were concluded with landlords whose main activity is outside the agricultural sector, against 16% only in Roi Et. Temporary or definitive emigration of rural labor to rapidly growing non-farm sectors near rural areas of Central Plain as well as speculative land investment in this area are mainly responsible for the development of a group of "non-farmers landlords".

Table 1.14: Landowners' Occupation and Landlord/Tenant Relationship

	Suphan Buri		Pitchit		Roi Et	
	Percent of Contract	Percent of rented land	Percent of Contract	Percent of rented land	Percent of Contract	Percent of rented land
Occupation of landlord						
Non agricultural	**51.9**	**49.2**	**36.1**	**34.2**	**15.8**	**16.9**
Commerce	24.1	23.9	19.4	14.2	10.5	14.6
Manufacturing	3.7	4.0	11.1	10.8	-	-
Services and transport	24.1	21.2	5.6	9.3	5.3	2.4
Agricultural	**55.6**	**50.8**	**66.7**	**65.8**	**84.2**	**83.1**
Landlord/tenant location						
Same district	46.6	20.6	42.5	34.3	84.6	54.0
Same province	20.5	18.9	22.5	3.7	-	20.0
Same region	12.3	24.4	7.5	24.1	3.8	-
Different regions	20.5	18.2	22.5	7.1	11.5	7.3
Bangkok	-	17.9	5.0	27.6	-	18.7
Landlord/tenant kinship						
Close relatives	19.2	10.1	31.9	12.7	7.1	0.7
Distant relatives	38.5	24.0	27.7	15.8	53.6	40.2
Neighbors	26.9	17.0	27.7	17.7	32.1	11.9
No relation	15.4	48.9	12.8	53.8	7.1	47.2

SOURCE: ORSTOM/CUSRI Project, Survey on Rice Farmers

Second, there are two main reasons to expect the tenancy market to be very local. Since land area is often split into different parcels separate from one another, leasing in and leasing out may take place among farmers within a village for reasons of convenience and efficiency. Furthermore, farmers might be reluctant to lease land outside of their village area mainly for fear of increasing supervisory costs, losses due to crop damages and theft, and because the cost of negotiating and enforcing tenancy contracts is likely to be high when people from different areas are involved. In contrast, the landlord who lives in the village can interfere with the tenants' farm management practices. Data on the location of the land owners compared to the location of the land (thus of the tenant) provide evidence that absentee landlords, defined as non-resident landlords, has developed in more urbanized areas. With the increased social and sectoral mobility, there emerged many landlords residing far away from their tenants. In Suphan Buri and Pitchit, nearly 30% of the owners live in a different province than the tenant that they are involved with, and about 20% in different regions. By contrast, in Roi Et, the market for tenancy has remained typically local, in the sense that the actual and potential participants are drawn from a limited area, usually the village in which the land is located.

Third, as far as landowner-tenant kinship is concerned, nearly 60% of the contracts were drawn up between relatives, irrespective of the province. The main reasons explaining this pattern is related to inheritance rules and imperfections in capital markets which make the purchase of land almost inaccessible. Because access to credit is difficult and land prices high, siblings who want to work the land usually rent in the other's holdings. The predominance of kinship among contractors tends also to support the hypothesis that returns to specific experience associated with land make intra-familial and inter-generational joint production of that land optimal. Transaction and monitoring costs tend to be lower if there are kinship ties or a tight social bond between landlord and tenant. For example, the closer the personal relationship, the greater the damage in the long run to a tenant farmer's reputation if he follows an opportunistic course.

1.3.4 Land Rentals

The arrangements under which land is rented define the conditions and thus the price of indirect access to land for those owning no land or too small pieces. In Thailand, there are two broad forms of tenancy contracts depending on the manner of rental payment. One type of contract consists of a proportion of produce to be paid as the rent, which therefore varies from season to season in accordance with output. This is usually referred to as share tenancy[14]. The second form of contract involves a fixed amount of rent either in cash or in kind, which is to be paid irrespective of the output level. This kind of contract is called fixed-rent tenancy. Whatever the rental agreement, the rent is ordinarily paid at rice harvest time. Rental-free agreements were observed in 12 cases. These contracts are mostly limited to very close relatives, usually between parents and children, in which parents provide a piece of land for their children to earn a living until the time of inheritance or purchase of land.

The system of renting land for a fixed amount is by far dominant among the sample of rice farmers studied; only seven farmers, located in Roi Et, reported to operate their land under share tenancy contracts. Several reasons explain the quasi disappearance of share tenancy. First, the change in the social background of landlords resulting from rapid industrialization since the 1980s, have weakened the advantages of sharecropping. Transaction costs under share tenancy tend to be high for both the landlord and the tenant if they live far apart. The former has to monitor the work of the latter from a distance, while the latter has to travel far for negotiations and to obtain inputs under cost-sharing arrangements (Fujimoto, 1996). The higher costs involved in implementing sharing arrangements have resulted in the spread of fixed rent tenancy.

Second, profit sharing tenancy is a means of capital mobilization under conditions where credit markets are underdeveloped. The profit-sharing system usually includes the extension of credit from landlords to tenants, because it is quite common for a landlord to be less credit-constrained than his tenant. This choice becomes meaningless as soon as financial constraints are resolved. Hence, the fixed cash (or in kind) system is chosen in more cases when financial constraints have been diminished as a result of easier access to the credit market or of increased cash income from nonfarm jobs. The

[14] One third to one half of the output is usually given to the landlord. For cases in our sample, we estimated the rent per rai according to the production level.

disappearance of share tenancy in Thailand must be thus related to the development of institutional sources of credit since 1975 (cf. chapter 4).

There is no "standardisation" of fixed-rent contracts. The amount of rent paid as well as the form of the payment, is a matter of mutual agreement between the respective partners. Within the category of fixed-rent contracts, an important distinction is whether the rent is paid in cash or in kind. As shown in table 1.15, while in Pitchit and Roi Et fixed rent in cash is prevalent, in Suphan Buri fixed rent in kind is more widespread. The level of agreed upon rent varies considerably from one province to another, reflecting various complex factors. The average rental in cash range from 560 baht in Suphan Buri to 344 baht in Pitchit and 250 bath in Roi Et. The average rental in kind amounts to 182 kg in Suphan Buri to 111 kg in Roi Et. Dry season rental rates are, on average, lower than rainy season rates.

Let us examine the factors that bring about these differences. Basically, the rent is expected to be positively related to the productive potential of the land; the higher the income from land use, the higher the contract fee paid for it. Improvements in land productivity through drainage and irrigation facilities resulted in higher returns to farmers in Suphan Buri and to a less extent in Pitchit. These higher returns are reflected in higher rents charged by landlords in Suphan Buri and Pitchit. Conversely, the main reason behind the low rental price for land in Roi Et can be attributed to low potential productivity of the land. The fact that the rainy season crop generate higher profit potential than the dry season crop due to better water availability also lead to a substantial premium in market rents.

The concentration index of land ownership can be a very crude index of the relative bargaining power of landowners and tenants; the larger the land concentration is, the stronger the bargaining power of the landlords is in the land market, and therefore the higher the land rentals will be. Judging from the variations in this index discussed in the preceding section, the ranking of the land rents is obviously consistent with that of the bargaining power of landowners. Where land concentration is high such as in Suphan Buri, land rents are high, and where land concentration is low, land rents are low such as in Roi Et.

Table 1.15: Average Land Rentals

	Kind (kg per rai)				Cash (baht per rai)			
	First Crop		Second Crop		First Crop		Second Crop	
	Frequency	Av. rent	Frequency	Av. rent	Frequency	Av. rent	Frequency	Av. rent
Suphan Buri	44	182	34	168	13	559	24	259
Farmer owner	22	189	18	158	6	585	9	328
Non-farmer owner	20	180	16	179	6	560	13	251
Relative	24	180	22	150	3	519	9	114
Non relative	20	186	15	179	10	571	15	346
Pitchit	3	125			27	344	5	158
Farmer owner	2	118			15	335	4	147
Non-farmer owner	1	140			9	372	1	200
Relative					15	295	3	159
Non relative	3	125			12	405	2	156
Roi Et	12	111			5	254		
Farmer owner	8	114			3	267		
Non-farmer owner	2	74			1	139		
Relative	5	96			4	235		
	7	122			1	333		

SOURCE: ORSTOM/CUSRI Project, Survey on Rice Farmers

Finally, as far as the owner's occupation is concerned, there is no evidence that non-farmer lessors exploit their tenants (who in many cases are relatives and friends) by charging rents above competitive norms compared to farmer lessors. As expected, agreed rentals are lowest under contracts established between close relatives and the highest in case of no relationship, especially when the rent is paid in cash. Better-off landowners tend to provide a lower rent to their relatives under the tradition of mutual assistance.

1.4 CONCLUSION

Thailand used to be a relatively land abundant country. The fact that over time new land has continuously been brought into cultivation helped both to reduce the incidence of poverty and to sustain agricultural growth. Few Third World countries have managed to maintain for so long a pace of land expansion that has surpassed the population growth so as to lead to an increase in the land-man ratio. Therefore, as population pressure has intensified neither the distribution nor the average size of holdings has greatly changed at the national level, and Thailand has succeeded in avoiding an increased incidence of landlessness.

Since the area currently under cultivation was once covered by forests, increases in rice production which have occurred during the past decades may be attributed to one major reason, deforestation. Because land was converted from forest cover to higher value uses, it has long been true that the highest social value of land use in Thailand was crop cultivation rather than forest cover except for very marginal areas. However, as the forest was further destroyed, the income loss from environmental damages started to outweigh the income gain from land expansion. Deforestation damages are cumulative and increased as the forest was further encroached, while the income gains from land expansion decreased. According to Panayotou and Parasuk (1990), the turning point has been reached, and further forest conversion to farmland might increase rather than reduce poverty over the long run.

Considering the rice sub-sector, the current distribution of land ownership does not support the hypothesis that rice land is highly concentrated in the hands of a few wealthy landlords. The conceptual couple land-owning/landless and exploiting/exploited do not adequately account for the relations between and within rural households. Rather, a continuous spectrum of peasant subclasses was found, holding various types and degrees of rights on the land they cultivate: there are households with only use rights, others with individual established ownership titles, still others with access to land only through contract farming schemes, and many with plots too small and infertile to secure subsistence.

The abundance of good, fertile and accessible land, explains the remarkably high rates of land ownership throughout the country. But equally obvious is the government policy which has put state lands at cultivators' disposal; land rights have always been given to farmers who had cleared and occupied land for a long time, even in the so-called reserved areas. The legalization of encroachers certainly contributed to weakening the control of landowners over the socially and economically weaker groups of the villages and the monopoly they tended to exert periodically on their labor (as tenants or sharecroppers).

Although rich farmers may have been acquiring more land for cultivation, there is no proof that land markets have strongly worked against small and marginal farmers rather than in favor of them. The evidence examined here is inconsistent with the opinion that rural conditions are gravely deteriorating and with the perception that small operators are losing their land to larger ones. Sellers were more frequently large landowners who sold only a portion of their holdings while most purchasers were agricultural entrepreneurs seeking to enlarge their farms. Transactions in land are not the main factor behind skewed land distribution. However, the closing land frontier is expected to deepen differentiation within the peasantry and is bound to transform land accumulation patterns into further concentration of land ownership. Since no virgin land is further available for cultivation the necessity of purchasing expensive land out of savings might tend to make the distribution of landholding more unequal; land is likely to accumulate faster in the hands of farmers with already large savings and thereby reinforcing large land ownership.

Tenancy is a common (but not pervasive) phenomenon in all the study areas. Opposite to a widespread opinion considering the relationship between landlord and tenant to be unequal and exploitative, there is much evidence that the modern landlord-tenant relationship has acquired a rather different character. Tenancy in rice growing areas appeared neither inherently nor inevitably damaging to the interests of the poor. It is argued that this is a socially useful market response which provides opportunities for a more efficient employment of family resources. The land rental market allowed households with differing demands for land to obtain the amount of land that best matched their endowment and preferences. As a result, the market for the rental of rice land tended to make operational holdings more equal and closer to optimal operational size than ownership holdings.

Chapter 2

RURAL LABOR MARKETS AND OFF-FARM EMPLOYMENT

2.0 INTRODUCTION

In developing economies, an important factor influencing a country's ability to absorb a growing labor force into productive employment is the type of strategy pursued for developing the agricultural sector. In Thailand, nearly 50% of the labor force still earns its livelihood from the agricultural sector, and despite the strong movement towards agricultural diversification since the 1960s, more than 60% of rural laborers are rice farmers. A major reason why rice farming has been able to retain most of the agricultural labor force has been both its ability to absorb high levels of labor input and a huge availability of land suitable for rice cultivation.

Although the closing of the land frontier during the 1980s induced a steadily declining supply of arable land, land to labor ratios did not change much in the rice sector. This is because meanwhile, Thailand exhibited growing economic prosperity, which brought dramatic increase in the non-farm demand for labor. With opportunities for more remunerative employment rising elsewhere, workers moved out of low-productivity, low wage rural economic activities. As a result, labor progressively became a key constraint to increasing rice production. Under these conditions, the potential that the rice farming sector holds in releasing labor force is of crucial importance. This, in turn, raises the important questions as to how labor is used in rice farming, how much labor is required, what labor is available, when is it available.

The pattern of land distribution and technological choices are two key determinants of the labor demand conditions within the rice sector. Farm labor use varies first according to farm size, labor intensity being much lower on large farms than on small ones. The distribution of farm sizes has thus, at the aggregate level, a significant impact on the demand of both family and hired labor. Because of its policy importance for land reform, for natural resources management and migration, the inverse relationship must be examined carefully. Numerous theoretical papers have shown how dualistic labor markets (with family labor cheaper than hired labor because of incentives problems) create incentives for larger landowners to use less labor, and small holders to be more intensive in labor. However, imperfect labor markets cannot alone account for a systematic inverse size-labor relationship. Differential intensity of labor utilization has to be explained in terms of a multiplicity of labor market imperfections, which means that the policy implications extend well beyond redistributive land reform.

Second, a farmer has a variety of agricultural techniques to choose from. Different techniques have different requirements for land, labor and capital. Traditionally, Thai rice farming techniques have been highly intensive in human labor as well as land using. However, the recent competing demand for land and labor in non-agricultural economic activities affected their relative scarcities and prices. As a result, there has been a tendency to choose both land and labor-saving techniques in the rice sector. Whether this has led to new employment creation or to employment destruction remains an open question, because the labor days saved due to the progress in mechanization can be more than compensated by higher labor use due to increased incidence of multiple cropping, introduction of improved varieties and improved cultivation practices.

Finally, although the majority of rice farmers live in rural areas with on-farm activities as their main occupation, most households also allocate time to wage employment. Multiple job-holding and temporary migrations of labor are known to be common in Thailand. This reflects either or both of two contrary and opposing processes at work, usually described as « *push* » and « *pull* » factors. « *Push* » factors are related to the limited capacity the rice sector has to generate adequate income, which is forcing people to rely on wage employment to earn their living. « *Pull* » factor consists in the availability of attractive off-farm opportunities. Explicit attention to the interdependencies between the farm and the nonfarm allocation of labor is important because the labor supply to rice activities may differ significantly depending upon the amount of the off-farm labor income different households can earn.

This Chapter attempts to describe the current situation of labor absorption in the Thai rice sector. Factors responsible for the variations in labor use for rice cultivation as well as the determinants of the involvement of household members in off-farm activities are considered. They are grouped into three kinds of variables: household characteristics such as members' age, sex and education; farm-level factors such as scale of farming, land tenure, and rice production technology; labor market-level factors such as the earnings of each activity.

2.1 Family Labor Force, Labor ExchangeAnd Hired Labor, Farm and Off-Farm Work: Definitional Issues

2.1.1 The Composition of the Rice Labor Force

There are three main sources of agricultural labor in Thailand: family labor, hired labor and exchanged labor from other households.

The family members include all individuals who where domiciled with the household (i.e., shared meals and used the same facilities) at the time of the survey[15]. The family labor force available to each household has then been measured in adult equivalents: according to the rules of the Labor Force Surveys in Thailand, any family member from seventeen to fifty-nine years of age were counted as one, irrespective of gender; elders (sixty and over) and

[15] A source of confusion was related to the status of some members who migrated to look for off-farm employment. Some of them were permanent migrants domiciled elsewhere whereas others were still domiciled with the household. Some full-time year-round outworkers continued to reside in their home village and travel to their place of work on a daily basis. They continued to participate in the family decision-making process and to remit a share of their income to other household members. They were unambiguously integrated as household members. Outworkers domiciliated somewhere else were unambiguously excluded from the household as well. For ambiguous individuals the final definition used to differentiate household from non-household members was drawn from the household heads themselves, that is, we included persons whom the heads considered as integral parts of the household.

adolescents (eleven to sixteen) were counted as one half; young children under eleven years old were excluded, that is counted as zero.

The second source of labor available to rice farmers is temporary workers. Casual laborers can be of different types. There are permanent «*professional*» migrant laborers who reside non-locally and who travel through the regions seeking wage work and temporary laborers (neighbors most of the time) whose yearly incomes are supplemented by the wage work they obtain locally during the planting and harvest season. Wage labor can involve individual or group contracts; in the latter case, a group of laborers headed by a leader concludes a contract with the farmer. Wage laborers are usually hired on a casual basis, be it daily or piecework in order to complete an operation lasting a few days. Usually, the laborer's relation with the employer ends as soon as the season is over.

The last source of labor for rice farming operations is exchanged labor, which is a form of traditional and reciprocal exchange of labor between neighboring households. The customary exchange agreement is that one household asking for hands from other households is obliged to return to them exactly the same amount of labor power, which is basically measured in person-days. Therefore, a household, which receives unpaid assistance from neighbors, accumulates obligations to help others.

2.1.2 Characteristics of Non-Family Labor and Labor Contracts

2.1.2.1 Characteristics of Non-Family Labor

Table 2.1 shows that the tradition of exchanged labor has almost disappeared in the areas undergoing technological change; in more traditional areas a good deal of labor is still employed under non-market relations and wage labor is only a fraction of total hired labor. Around 20% of the households surveyed in Suphan Buri and Pitchit called on exchanged labor against 60% in Roi Et.

According to responses from farmers, there are two main reasons for the dying out of labor exchange. In the irrigated areas, the most important one is related to technological progress. As the flexibility in timing agricultural activities with double cropping lessened, the need for extra labor increased for every household at the same time, making labor exchange impossible. The second reason is linked to the marked preference of workers for remuneration in cash rather than in kind. Hence, hired labor appears principally as a substitute for exchanged labor, which has generally occurred under the penetration of the money economy and the development of capitalistic relations of production in the rural sector[16]. The method for supplementing family labor during peak seasons, i.e. whether to hire workers or set up reciprocal relations with neighbors, is thus determined by socioeconomic relations between villagers, which in turn have had a lot to do with the degree to which the market economy has permeated the community.

Table 2.1: Recruitment and Origin of Workers

	Suphan Buri	Pitchit	Roi Et
Percentage of households using exchanged labor	17.0	23.0	61.0
Percent of households resorting to			
personal recruitment	34.7	89.4	96.0
middlemen recruitment	65.3	10.6	4.0
Percent of households whose workers come from:			
Same district	34.8	84.2	83.3
Same province	40.0	12.6	16.7
Same region	20.9	3.2	
Other region	4.3		

SOURCE: ORSTOM/CUSRI Project, Survey on Rice Farmers

Table 2.1 also shows that labor contracts are still very often concluded within a district community or within adjacent districts. The geographical mobility of workers employed by rice farmers is limited, especially in Roi Et and Pitchit where more than 80% of households called on casual labor within the same district. In Suphan Buri the labor market tends to grow beyond the boundary of a district: 40% of hired laborers came from other districts and 20% of hired laborers came from other provinces. This is because the increase in cropping intensity, made possible by increased availability of controlled water supplies, led to a sizable growth in farm labor inputs in this province, as

[16] In the further presentation of the results on non-family labor inputs no distinction has been made between exchanged labor and wage labor. Hired labor refers to the one being used for others regardless of the nature of the transaction, be it traditional or modern.

well as a decline in the flexibility of timing. Therefore, farms are surrounded by other farm families who also require large amounts of labor at the same time. They all have to go outside the community to find workers.

Consistent with the distribution of workers' origin is the method of recruitment. While in Pitchit and in Roi Et most households recruit workers by themselves, 65% of farms in Suphan Buri call on intermediaries. Established relationships with the local laborers, which in many cases are neighbors and sometimes friends, are open, informal and diffuse compared to the relationships with migrant workers. On the same way, given the reciprocal nature of exchanged labor, it necessarily takes place between neighboring households. In both cases, « *buyers* » and « *sellers* » of labor are known to each other sometimes for generations and there is no need for any kind of intermediaries to discuss the labor exchange agreement.

2.1.2.2 Labor Contracts

Standard economic theory assumes that the co-existence of seasonal fluctuations in income and labor demand should involve tied contracts to dominate labor markets. Under such conditions, the *patron-client* ties are mainly motivated by the need to insure oneself against unforeseen or random fluctuations, such as the presence of uncertainty in labor demand for the *patron* or fluctuations in income for the *client*. However, the empirical literature on rural labor markets clearly suggests that tied labor is far from being widespread (Mukherjee and Ray, 1995; Platteau, 1995). Empirical observations from our field survey point to the absence of long-term attached laborers for four main reasons.

A high level of seasonality in the labor market activity, combined with agricultural practices that make time constraints tight should imply a high proportion of tied labor. As the technology of rice farming involves steep fluctuations in labor use, there is scope to tie labor. On the one side, in the peak season many agricultural operations have to be performed in a very short interval, and there are risks and costs in delays. In such a situation, the employer should be interested in some commitment on the part of laborers regarding the timing and speed at which a given task is performed. On the other side, tied laborers need to be provided with a minimum amount of work and, therefore, earnings all year round. This is costly to the employer and it is possible for the losses incurred during the slack season to balance the gains of secure access to labor in the peak season. The final balance critically depends on wage differentials between the slack and peak season. This differential can be very wide, especially in rainfed areas (cf. infra); starting from a very low level, agricultural daily wages increase twofold to threefold between the two periods. By contrast, in irrigated areas, slack season agricultural daily wages are only 20-30% lower than those of peak season, making permanent laborers very expensive.

However, the fundamental reason why permanent laborers are not observed in the rice farming sector of Thailand may not so much lies in an unfavorable cost/benefit ratio than in an unavailability of laborers during the lean season. There is a rationale for labor tying arrangements mutually desired by both employers and workers only if labor shortages during the peak season coexist with a supply of excess labor during the slack period. Indeed, the seasonality of rice production implies a significant change in the utilization of labor over the year. The rice labor force expands in the rainy season in response to heavy demand for labor in on-farm activities, whereas it contracts during the dry season. This contraction doesn't mean that rural laborers withdraw from the labor market. Under the impact of industrialization, rural laborers have had access to an increasing number of alternative employment opportunities during the slack season. The fact that wage differentials between agricultural and non-agricultural jobs reach their maximum during the slack season certainly caused the bargaining power of rural laborers to increase. In such a setting, employers have tended to substitute capital for labor provided that labor-saving technologies were available, while laborers have tended to seek alternative sources of income outside the agricultural sector.

Another rationale may exist for long-term labor contracts stemming from the existence of care-intensive farm tasks, among which worker supervision. If a general abundance of unskilled labor coexists with a scarcity of a specific type of worker, the institution of permanent labor performs a useful function by economizing on supervision costs when agricultural tasks are inherently difficult to monitor. In this strategy, employers try to attract those laborers who can be expected to reliably perform care-intensive farm tasks. From this point of view, the employment of regular farm laborers has not been attractive for both small and large rice farms. For very small farms relying mostly on family labor, the supervision constraint, i.e. the necessity to employ permanent labor, is less binding. The large farms in Thailand have reduced the importance of labor supervision through investment in the mechanization of cultural operations. One of the critical advantages of mechanization is that it often implies that the pace and quality of output are primarily determined by machine specifications rather than by the quality of the operator's effort.

Lastly, tenancy contracts may be considered as an available alternative to permanent labor inasmuch as tenancy is socially and legally accepted. Permanent labor contracts may be optimal only if the spectrum of contractual choices is institutionally or socially restricted. In Thailand, large landowners have obviously found it more profitable to rent out their land than to cultivate it with a large number of regular farm laborers.

2.1.3 Farm, Non-Farm and Off-Farm Activities

The data gathered during the field survey refers to the time spent on every activity, roughly measured by the number of days. There is thus little ambiguity regarding intensity of labor utilization as there may be with data on numbers of workers, and comparisons between households are relevant[17].

We have then defined four broad categories of time use. Total time endowment of each household member can thus be used for farm work, non-farm work, off-farm work and home time. Although off-farm activities have received considerable attention in the literature and among policy makers, they usually are a source of conceptual confusion. The main source of the confusion comes from the fact that off-farm employment is often viewed as work of a nonfarming nature and is therefore considered to be non-farm employment. This view results in very misleading conclusions when working members of rural households hold jobs away from the farm in the agricultural sector. In our study, work is classified according to two criteria, i.e. the branch of industry and the location of the job: on-farm production of agricultural products operated and managed by the household is farm work, on-farm production of non-agricultural products is non-farm work.

Therefore, the definition of economic activities for which days of work were reported in the survey was limited to three major categories of work.

[17] One shortcoming of the survey comes from the fact that one family member provided information for the whole family; in most cases it was the head of the household. Because the head sometimes had difficulty telling how much time each member of the family spent on each activity, interviewers asked ordered qualitative questions, that is if any worker was considered to work one third of a full-time job, up to two thirds, full-time, or not at all. In all these cases, survey interviewers prevented people from indicating that any worker worked full-time both on and off farm.

Farm activities of the household include rice production activity, other agricultural production activity such as vegetable and fruit farming, livestock and poultry, fish and shrimp. Non-farm enterprises include a wide range of activities such as commerce, silk and cotton weaving, pottery, wood and bamboo products, etc. Off-farm work includes all work in agricultural as well as non-agricultural enterprises owned or operated by other households or firms. Non-market activities or personal activities such as cleaning the house, cooking, child care, other household maintenance activities, general maintenance of equipment, and animal feeding for home consumption have not been covered by the survey. This omission led to an underestimation of total time spent on work, especially for women and teenagers.

2.2 FAMILY DEMOGRAPHICS AND THE HOUSEHOLD SUPPLY OF LABOR

Choosing an occupation rather than leisure consists of a number of inter-related decisions. From an individual's point of view these are: whether to work and if yes, how many activities to undertake and the nature of the economic activities (agricultural/nonagricultural, on-farm/off-farm, self/wage employment). As participation in the labor market depends on the personal and family characteristics of workers, the relevant parameters are family size, age and sex composition of the household labor force, and human capital accumulation.

Among the several factors that determine the use of family labor, the most obvious one is its availability. Table 2.2 presents the patterns of time allocated by family members in farm, non-farm and off-farm activities as well as the use of hired labor in rice farming according to the size of the family labor force. A brief look at this table reveals that family members are the prevalent source of labor time in rice farming, and rice farming is the households' major occupation: the total work spent on-farm activities is around 70-90% of the total time worked, of which around 60-80% is spent on rice production only. In Pitchit household agricultural activities are more diversified and, consequently, family workers are more engaged in activities other than rice, which represent around 20% of total labor time. Comparing labor allocation between family size groups, it can be noted that rice labor requirements as a share of total labor decreases with family size, whereas the time devoted to other agricultural activities increases.

Table 2.2 also documents a widely observed phenomenon, i.e., increasing family size is associated with higher involvement in off-farm activities. The higher the number of working adults in the household, the more labor days there are available for hiring out. In Suphan Buri, the smallest households spent nearly 12% of total working time on off-farm activities while larger ones spent nearly 30%. The respective figures are 8% and 11% in Pitchit and 11% and 13% in Roi Et.

It is generally understood that casual labor is complementary to family labor during peak seasons while it is often regarded as a substitute during slack seasons. Farmers with little family labor available have the greatest incentive to hire workers whereas the greater the number of farm workers in the family, the lower the incentive to use casual laborers. Data in table 2.2 confirms that use of casual labor correlates negatively with the size of the permanent labor force, that is the share of hired labor in total labor use for rice farming decreases with the size of the household.

The pattern of labor allocation of rice households is largely based on family composition. The structure of the households and the related household composition of labor supply are given in Table 2.3. The number of family members in different age groups and sex affects total labor supply and labor allocation in various ways. There are noticeable differences in the time-allocation pattern of men and women. It has often been argued that cultural and status constraints can prevent women from offering themselves on the labor market. Women may be reluctant to take up regular outside work because of interference with household tasks and the responsibility of child caring. In contrast, males may have a comparative advantage in off-farm work because of their ability to travel farther away from home and earn higher wages. The data suggest that females spend more of their time on rice cultivation and on nonfarm activities than males while they participate less in off-farm work.

Table 2.2: Household Labor Force and Labor Supply

	Suphan Buri			Pitchit			Roi Et		
Number of workers	1-2	3-4	5-6	1-2	3-4	5-6	1-2	3-4	5-6
Labor input on rice production									
Family worker - days per rai	27.2	32.4	34.3	26.7	22.2	23.4	37.9	67.8	27.8
Hired worker – per rai	5.7	5.5	5.5	4.4	3.3	3.0	5.1	2.5	6.98
Hired labor for rice in % of total labor	20.3	17.7	13.4	22.0	16.6	15.4	23.4	7.9	19.5
Labor allocation (in %)									
Rice cultivation	72.8	64.3	57.2	72.1	72.8	67.0	74.7	83.5	74.7
Other agricultural activities	6.0	12.5	10.8	19.6	20.8	21.8	0.9	0.5	6.6
Non-farm activities	9.4	10.7	4.2	-	-	-	13.8	3.3	5.7
Off-farm labor	11.7	12.4	27.8	8.3	6.4	11.2	10.7	12.7	13.0

SOURCE: ORSTOM/CUSRI Project, Survey on Rice Farmers

Table 2.3: Labor Allocation (in %) According to the Workers' Age Group and Sex

	Males						Females					
	11-19	20-29	32-39	40-49	50-65	All Ages	11-19	20-29	32-39	40-49	50-65	All Ages
Suphan Buri												
Rice cultivation	60.1	67.6	57.8	68.3	72.4	66.7	57.0	64.3	63.5	75.4	64.7	65.8
Other agricultural activities	9.1	2.3	20.1	5.9	10.8	9.4	-	9.4	14.3	0.3	19.9	10.8
Non-farm activities	13.7	4.6	-	10.4	10.4	7.7	22.7	10.2	10.7	21.0	14.6	14.4
Off-farm labor	17.1	25.5	22.1	15.4	6.4	16.2	20.2	16.1	11.6	3.3	0.8	8.9
Pitchit												
Rice cultivation	68.2	70.3	79.6	64.4	74.4	71.9	68.9	79.7	74.5	70.7	80.3	75.8
Other agricultural activities	19.8	25.2	7.9	29.8	23.2	22.2	20.6	9.5	15.8	18.9	14.7	15.2
Non-farm activities	-	-	-	-	-	-	-	-	-	-	-	-
Off-farm labor	12.0	4.5	12.4	5.8	2.4	5.9	10.5	10.7	9.7	10.4	5.1	9.0
Roi Et												
Rice cultivation	77.6	77.6	69.2	75.7	88.6	77.5	80.5	82.1	71.7	89.4	92.3	81.4
Other agricultural activities	1.3	1.6	1.5	1.2	1.2	1.3	-	1.1	1.9	-	1.5	1.0
Non-farm activities	-	-	9.8	6.6	2.0	3.9	2.0	1.1	19.6	10.6	6.2	8.8
Off-farm labor	21.1	21.1	19.5	16.5	8.2	17.2	17.5	15.6	6.8	-	-	8.8

SOURCE: ORSTOM/CUSRI Project, Survey on Rice Farmers

Table 2.3 provides additional information on time allocation patterns of men and women by age group. The figures reveal persistent differences in the time use irrespective of age category: women are more likely participants in rice and non-farm activities, whereas men are more likely to participate in labor market activities and agricultural activities other than rice.

Age is known to affect labor supply in a nonlinear fashion: labor supply first increases and then decreases with age. The age of the individual is also expected to be important in the off-farm migration and the theory suggests a higher propensity to migrate for the young than for the old. One reason is that changing occupations and changing sectors, even temporarily, is costly. The cost of migration may be lower for younger workers than for older ones, especially when younger workers do not support additional family members. According to data in Table 2.3, outworkers are young. The allocation of time to activities other than farm production tends to increase up to the 30-39 year age bracket and decreases afterwards. Off-farm labor supply of both men and women reaches a maximum at the 11-19 or 20-29 age group depending on the sex (the peak is earlier for females than for males). In contrast, as workers get older they spend less time on off-farm work: workers in the 50-65 age group usually spend less than 10%, if any, of their total labor time on off-farm activities.

An interesting argument often put forward to explain differences between households' labor supply is related to the concept of the dependency ratio. The dependency ratio is given by the number of household members per adult equivalent family worker. The presence of family members in certain age groups classified as dependents[18] is expected to decrease total labor supply because of lower family labor endowments while family members in other age groups would shift labor supply in the opposite direction. According to data in Table 2.4, household labor input is negatively related to the dependency ratio: the more dependents per worker, the less the total household labor supply.

As far as labor allocation is concerned, it is questionable whether the presence of dependents should reduce or increase the time allocated to off-farm activities. On the one hand, for a given number of farm workers, the need to produce more per worker should increase as the number of household

[18] In our survey those under eleven years and those over sixty five years of age were both considered being dependents.

members that each worker must feed rises. One thus would expect labor allocation to rice farming to increase with the number of dependents. In addition, dependents require home time. The home-time devoted to dependents is likely to reduce the market supply of labor away form home. On the other hand, because off-farm employers usually offer better wages, the family's economic well-being is increased by mixed strategies of both farm and off-farm work. The need to earn more per worker as the dependency ratio increase might act as a strong incitement to off-farm labor.

We found a positive relationship between the number of dependents per family worker and the involvement of households in off-farm activities, except in Roi Et where both variables seem to behave independently. In Suphan Buri and Pitchit, the supply of off-farm labor is positively associated with the dependency ratio, which means that family workers burdened with a large number of dependents tend to work less intensively on their own farms and spend more time in off-farm activities in order to meet higher living expenses.

Finally, education is expected to increase both farm and off-farm labor supply because it has a positive effect on-farm work efficiency and also because higher wages are paid to more educated workers. Unfortunately, we have not been able to explore the relationship between education and labor supply and between education and activity choice for all household members because only the level of educational achievement of the household heads were recorded during the survey. In addition, we found a high homogeny in the level of education attained: nearly 95% of the heads were found to have completed four years of education. Formal education can therefore reveal little about ability of education to enhance an individual's performance in his chosen activity.

Table 2.4: Household Labor Allocation (in %) and Dependency Ratio

	d = 0	0 < d < 0.5	0.5 < d < 1	1 < d < 1.5	1.5 < d < 2	d > 2
Suphan Buri						
Total household workdays	957	1019	949	631	532	455
Labor allocation (%)						
Rice cultivation	59.9	70.3	66.5	69.9	64.1	70.1
Other agricultural activities	12.6	7.2	17.1	2.1	5.0	1.6
Non-farm activities	13.8	9.9	8.5	8.1	0.6	13.2
Off-farm labor	13.7	12.5	8.0	19.9	30.3	15.1
Pitchit						
Total household workdays	946	1165	886	691	711	483
Labor allocation (%)						
Rice cultivation	68.1	70.0	75.3	77.5	67.2	76.8
Other agricultural activities	25.7	22.8	15.6	18.2	19.1	10.9
Non-farm activities	-	-	-	-	-	-
Off-farm labor	6.2	7.3	9.1	4.2	13.6	12.3
Roi Et						
Total household workdays	671	998	809	710	541	499
Labor allocation (%)						
Rice cultivation	83.7	78.8	81.0	79.9	89.8	65.4
Other agricultural activities	1.7	4.1	1.0	-	-	-
Non-farm activities	2.6	3.6	9.2	5.4	-	24.0
Off-farm labor	12.1	13.5	8.9	14.7	10.2	10.6

SOURCE: ORSTOM/CUSRI Project, Survey on Rice Farmers

2.3 Scale of Farming, Land Tenure and Labor Intensity

The question of how the labor supply of workers from different economic groups is allocated is of considerable importance for the analysis of employment issues. The labor supply behavior among household categories may differ owing to the set of constraints they face depending on their economic status and asset ownership. The differences in the nature of constraints households face may introduce significant differences in the quantity of work performed in rice farming activities; time allocated to alternative activities is expected to differ significantly as well. An examination of the patterns that were observed in family labor use according to scale of farming and land tenure is included in this section.

2.3.1 The Relation between the Scale of Farming and Labor Use

The data in Table 2.5 shows both family and hired labor input per rai declining as holding size increases, so that the larger the farm the less labor intensive is its mode of production. This holds true within each province. The labor input in rice cultivation drops by more than 13-fold for households cultivating more than 100 rai in Suphan Buri, by 7-fold in Pitchit and Roi Et. A related phenomenon to declining labor inputs per rai of cultivated rice is the declining ratio of family labor to total labor, as farm size increases. In Suphan Buri and Roi Et, hired labor ratios reach a peak for the middle-size farms and are comparatively lower for both the smaller and the larger holdings. In Pitchit, hired labor adds, on average, 36% to the work done by the family of the largest farms and 6% to the work done by the family of the smallest ones.

Table 2.5: Total Labor Input by Landholding Size

	< 15 rai	16-30 rai	31-50 rai	51-100 rai	101+ rai	All sizes
Suphan Buri						
Labor input on rice production						
Family worker-days per rai	40.9	29.9	16.7	16.1	3.0	29.8
Hired worker-days per rai	7.1	5.7	4.9	3.5	0.5	5.8
Hired labor for rice in % of total labor	13.7	18.4	21.9	19.6	14.0	18.3
Family labor allocation (%)						
Rice cultivation	62.1	69.1	64.1	70.8	100.0	66.4
Other agricultural activities	7.8	6.4	16.7	26.5	0.0	10.3
Non-farm activities	16.7	7.9	8.5	0.0	0.0	9.7
Off-farm labor	13.4	16.7	10.7	2.7	0.0	13.6
Pitchit						
Labor input on rice production						
Family worker-days per rai	46.0	29.3	17.5	14.5	6.6	24.9
Hired worker-days per rai	3.1	3.9	4.4	4.3	3.7	4.0
Hired labor for rice in % of total labor	6.1	12.0	19.9	23.5	35.6	18.7
Family labor allocation (%)						
Rice cultivation	71.1	78.7	70.3	68.7	65.2	71.5
Other agricultural activities	11.1	16.3	18.8	27.9	32.1	20.5
Non-farm activities	0.0	0.0	0.0	0.0	0.0	0.0
Off-farm labor	17.9	5.0	10.9	3.4	2.7	8.0

Roi Et						
Labor input on rice production						
Family worker-days per rai	74.4	29.2	19.0		10.3	53.4
Hired worker-days per rai	7.4	9.2	6.3		1.4	7.4
Hired labor for rice in % of total labor	4.9	18.4	23.9		12.1	13.6
Family labor allocation (%)						
Rice cultivation	83.1	76.6	77.5		87.8	80.6
Other agricultural activities	0.0	2.3	2.5		0.0	1.1
Non-farm activities	2.7	11.7	9.3		0.0	6.1
Off-farm labor	14.2	9.3	10.7		12.2	12.2

SOURCE: ORSTOM/CUSRI Project, Survey on Rice Farmers

Given the fact that the labor intensity of cultivation of individual holdings depends on the scale of their production, the average labor intensity of regions or provinces as a whole must depend on the size distribution of their farms. We saw in Chapter one that while the distribution of Pitchit has more cases toward the right end, Roi Et's has a bias toward smaller values, i.e. the distribution is negatively skewed or skewed to the left. The distribution of Suphan Buri is more symmetric around the mean. The regional differences in labor use are quite substantial and consistent with land distribution within each region. The highest per rai labor inputs were recorded in the Northeastern province and the lowest in the Northern province. Suphan Buri in the Central Plain occupies a middle position but is closer to the levels obtained in the North. If we compare the size-specific rates of labor input per rai of rice between the three provinces in Table 2.5, we find that households which fall in the same size group tend to use roughly the same amount of labor, except for workers belonging to the smallest farm holding group in Roi Et who spend more time on rice work than in the other provinces[19]. Hence, it appears that average family labor absorption variations across provinces are primarily affected through differences in size distribution of farms rather than through the size-specific rates of labor use.

The debate over the inverse size-labor relationship has been extensive and focused on four main areas: first, this relationship might be caused by differences in land quality, and thus cropping patterns and cropping intensities; second, it would be restricted to farms facing distress conditions; third, labor markets imperfections would be mainly responsible for the phenomenon; fourth, multiplicity of market failure creates incentives for larger farmers to apply a lower intensity of labor. The remaining part of this section discusses the validity of these arguments regarding rice growing areas in Thailand.

2.3.1.1 Land Quality and Cropping Patterns

A first obvious explanation of the higher per rai labor inputs into some holdings is that these holdings have a better quality of land, which is more profitable to cultivate intensively. It appears crucial, therefore, to take in

---

[19] The apparent specificity of Roi Et arises from a much greater number of landholding of less than five rai in this province.

account land quality. A first problem is to measure the quality of land. Ideally, a comprehensive description of the plot should be used, including measures of the topography of the plot, its location, and a categorization by local soil type[20]. Unfortunately, our data set does not include direct observations of land quality. Hence, indirect method has to be applied.

Irrigation is an interesting proxy for the productive potential of land: better irrigation leads to greater soil fertility and irrigation is usually considered as a precondition for multiple cropping which, in turn, is regarded as the most effective way of raising labor inputs per unit of land. Higher cropping intensities[21] should result with more irrigation. The tendency for irrigated land to be concentrated in the smaller holdings is supported by the data in Table 2.6.

Table 2.6: Land Fertility, Cropping Intensity by Farm Size

	< 15 rai	16-30 rai	31-50 rai	51-100 rai	101+ rai	All sizes
Suphan Buri						
Percent of irrigated area	99.2	96.7	98.4	86.3	50.0	97.1
Cropping intensity	156	171	152	150	100	160
Pitchit						
Percent of irrigated area	74.2	72.6	69.1	52.6	65.4	67.4
Cropping intensity	123	130	114	106	104	117
Roi Et						
Percent of irrigated area	0.0	0.0	5.0		0.0	1.0
Cropping intensity	100	100	100	100	100	100

SOURCE: ORSTOM/CUSRI Project, Survey on Rice Farmers

The smallest holdings in Suphan Buri are almost fully irrigated while the largest ones have only 50% of their land irrigated. The corresponding figures for Pitchit are 74% and 65% respectively, whereas all households of Roi Et, whatever the size of the holding, cultivate rainfed rice. Accordingly, cropping

[20] However, even with this quite rich description, some unobserved variations in plot characteristics might remain.

[21] Cropping intensity is the number of times the same piece of land is cultivated during the year. If water is not available for the dry season, the land is left uncultivated, resulting in a cropping intensity of 100.

intensity in the rice areas shows a tendency to decline with larger farm sizes once the size of 16-30 rai is reached. For farms of less than 15 rai, the intensity of rice cropping is lower despite higher percentages of irrigated area.

Second, even without being well irrigated, rice labor input into small farms may be higher because small operators adopt a more labor intensive cropping pattern. This pattern is likely to take at least two forms: more focus on crops using little land and much labor, and increasing intensity of cultivation of these products. In Suphan Buri and Pitchit where there are many opportunities in the choice of what to grow, part of the labor is shifted from rice labor to other agricultural activities in large farms (cf. Table 2.5). Decreases in rice labor intensity in the largest farm sizes are thus partly due to increases in labor devoted to other agricultural activities. In Roi Et, where these opportunities are almost non-existent, except livestock and poultry, which are not very, labor consuming, most of the work is devoted to rice in all farm size groups.

However, factors such as double cropping, a higher percentage of land irrigated, and focus on labor-intensive crops are not exogenous factors, they reflect the economics of farming. Therefore, differences in cropping patterns and cropping intensities simply pose the problem in another way in the sense that the central question (why do the smaller farmers « *choose* » to use more labor than the large cultivators) still remains unanswered.

2.3.1.2 The Subsistence Income

Among the factors that drive a small farmer to more intensive efforts, the most important one may be his need for survival. There is a certain basic minimum of consumption that a poor peasant family has to have without which it cannot survive. Such a poor peasant family, depending on a small piece of land (and having no alternative sources of employment and income) is pushed by distress to apply labor more intensively in order to secure supplemental output for survival. Here, the primary motivation is output maximization, which imply a high intensity in the application of all factors of production, particularly labor. Some authors have argued that it is precisely the intensive use of labor that has permitted the small farms to survive in the face of land and capital constraints (Berry and Clive, 1979). In other words, it would be somewhat misleading to treat the phenomenon as one of scale: this phenomenon would be restricted to the small farm size groups that are subject to distress conditions.

The first most obvious problem with this argument is that distress conditions are hard to assess. Second, there is the objection that a strong division between farms supposed to face distress conditions (small size, renting land, burdened with high dependency ratios) and the other ones was not observed. The inverse relationship between farm size and labor input appears in the data as a law that holds for all values of farm size.

2.3.1.3 Market Imperfections

Market imperfections have often been put forward to argue that small and large farms face a different set of factor prices, such that the effective price of land and capital is low for the latter and the effective price of labor is low for the former. The relatively low implicit price of land and capital for large farmers compared to the relatively low implicit price of labor for small farms would dictate choices of technique with different factor intensities; small farms would prefer high labor/land ratios, whereas large farms would choose to use labor less intensively. However, that factor markets are imperfect is controversial because it asserts that small and large farmers face different input prices and the proof of such an assertion is difficult. Let us examine here empirical evidence on imperfections in labor markets in Thailand.

The debate over the labor market imperfections has been constituted predominantly by two theoretical approaches. First, the dual labor market hypothesis developed by Sen arises from the alleged existence of two sectors within agriculture, the « *modern sector* » hiring workers and maximizing profits while the « *traditional sector* » uses mainly family labor and maximizes utility in a trade-off between output and leisure. To the farmer hiring in labor, the wage worker is a variable factor to be used up to the point where profit is maximized, i.e. where the marginal product of labor is equal to the real wage rate. To the small farmer, family labor is a fixed factor to be used to the point where its marginal productivity equals zero. Such a dualism leads to different shadow prices for labor between small « *traditional* » farmers who use mainly family labor and the large « *modern* » farmers who rely more on hired labor. Hence, family labor in small farms is valued at a lower implicit wage than that prevailing in the labor market, resulting in a more intensive application of labor on small farms.

A first criticism of the cheap family labor argument for the more intensive use of labor on the small farm in Thailand is the fact that small farms themselves hire labor and derive income from employment of family members in other occupations. The survey data indicate substantial purchases and sales of labor by all farms regardless of the size. Under such conditions, for the cheap labor argument still to hold it must be demonstrated that small farms value the time spent on rice farming by family labor at a lower price than the hired labor for equivalent tasks and/or that alternative employment opportunities are limited for small farms. The empirical evidence gathered in the study fail to support both assertions.

If family labor is cheaper than hired labor, then why do small farmers who have far more family labor available per rai than larger farmers hire in further expensive casual labor? The main reason is that rice cultivation requires supplementary labor resources regardless of farm size, necessitated by the ecology of rice cultivation itself. Hired labor is used to complement family labor in the critical periods of both the rainy and dry seasons when family labor is inadequate to perform all the required work within a limited time (cf. infra). Time constraints necessitate the hire of labor for urgent tasks to avoid harvest failures. Family and hired labor appear thus as complements, not substitutes. In such a case, the hypothetical gap between the marginal productivity of family labor and the market wage might be a statistical illusion or artifact caused by mismeasuring the agricultural wage rate. Since a substantial fraction of hired labor is employed in the peak season, the agricultural market wage rate is likely to predominantly reflect the effect of high peak season wage rates, whereas the marginal productivity of family labor should significantly include the depressing effect of the low productivity of family labor in the slack season. Therefore, it is possible for the average marginal product of family labor to be lower than the prevailing market wage, without meaning that the labor market is dual.

The other major argument we can oppose to the cheap labor hypothesis is that many operators of small holdings take advantage of non-farm activities or outside employment in the slack season, when non-agricultural casual wages are far higher than agricultural ones[22]. Most family workers of small holdings

[22] Some peasants even take these opportunities during the peak season and hire workers to take care of their farms. However, very few household members fall in this category. Most family

spend more time on non-farm and off-farm activities than workers of large holdings. In Pitchit, time worked off-farm increases from less than 3% of total time worked in large farms to nearly 20% in small ones. In Suphan Buri, non-farm work accounts for about 17% of total work and wage activities accounts for 13.4% of total work for the smallest farms compared to zero for the largest ones. In Roi Et the relationship between farm size and off-farm activities is not as clear-cut as in the other provinces: household involvement in off-farm activities does not vary significantly in the smaller sizes groups compared to the larger ones. On the whole, these high participation rates in off-farm employment suggest the absence of employment constraints in the rural labor market that prevent small farms from equating the marginal returns across activities, or in other words, from allocating their time up to the point where the marginal returns from farm and off-farm activities are equal to the market wages.

A second attempt to explain the inverse relationship between farm size and labor use, nested in the imperfect information paradigm, has been formulated in terms of qualitative factor differences between farm-size categories with respect to management and labor quality. Supervision problems and incentives become more problematic as size increases (Feder, 1985; Dyer, 1997). Hired labor would not exercise as much care and effort as family members, and therefore require supervision. This means that family time and effort has to be spent on supervision and monitoring. As holding size increases, the costs of supervision are likely to rise more than the marginal return of hired labor, especially if a hierarchy of paid supervisors needs to be instituted. In addition, there are search and hiring costs associated with outside laborers. This would create strong incentives for larger landowners to operate their holdings with a low intensity of labor. Small farms, by comparison, who rely mainly family labor, would have a productivity advantage with respect to the quality of labor. This is because farm families members, unlike hired labor, share the burden of risk and would be thus well-motivated, they do not bear search or recruitment costs in the labor market, they benefit from supervising each other, and they often provide non-marketable factors such as managerial ability. This in turn result in an intensification of labor use on small family farms.

---

members withdraw from rice cultivation during the slack season so that off-farm and farm work does not overlap and the labor allocated to rice generally do not compete with labor allocated to off-farm activities.

However, although these arguments seem to have considerable force, they would appear to have more relevance for vast landed estates or extensive ranch type-farming. They have much less relevance for the types of large farms we are considering, in the range from 50 to 100 rai. Under such conditions, it is likely that supervisions costs and managerial diseconomies are exaggerated. In addition, the fact that in a great number of contracts, farmers and workers live in the same village or usually have kinship ties help to provide efficient supervision of labor.

Overall, the thesis of differential labor prices between large and small farms resulting from dual labor markets or from supervision constraint does not appear to be relevant in Thailand. Therefore, the usual dichotomy of households by objective function (small farm households maximizing utility, large owners maximizing profits and using only hired labor) appears to be not only counterfactual but less useful than distinguishing large and small farms according to whether they are net importers or exporters of labor services. Competing explanations of the inverse relationship must be taken seriously.

More recent work has highlighted the fact that imperfections in capital and financial markets also explain why large farms are more likely to adopt technologies requiring fixed capital than small farmers are (Heltberg, 1998). Institutions and services (extension, credit, information, marketing) and, as a result input/output prices, exhibit a strong bias in favor of large farmers. The survey data supports the proposition of capital price differentials as the main explanatory mechanism at work. Small holdings have limited access to any kind of credit and display a high occurrence of loans from non-institutional lenders, who charge higher interest rates, give smaller loan amounts and for shorter periods than banks and cooperatives (cf. Chapter 4). Large farmers, by contrast, disproportionately benefit from institutional credit. Since the price of capital is partly fixed by the interest rate charged by lenders, the small farms lose advantage vis-à-vis larger ones, which have better access to cheaper credit. This in turn tends to encourage the substitution of capital equipment for labor on large farms. Accordingly, field data show higher capital intensities on large farms manifested in labor saving innovations (cf. Chapter 3).

2.3.2 The Relation between Land Tenure and Labor Use

Since access to land is the most important determinant of on-farm employment, land tenure is expected to affect the extent to which farm household members participate in farm activities. The common prediction of economic theory is that farmers apply themselves (and other inputs) more on their own land than on land leased in, because owners are given incentives to use land more intensively. Figures in Table 2.7 show that, as far as family labor is concerned, this relation does not seem to hold. For example, owner farm operators in Suphan Buri reported rice cultivation time of around 35 mandays per rai and per year, whereas tenants spent approximately 32 mandays. In Pitchit, tenants have the highest intensity of labor use.

These results are not terribly surprising. For an agricultural household, transactions in the land lease market are not made in isolation of its endowment structure and its decisions in other markets like those of labor and capital. It is therefore to be expected that the extent of tenancy is influenced by the transactions in these latter two markets, as well as by the family size and composition, and the ownership of capital stock. We have already shown in Chapter 1 that tenancy is an essential mechanism for the adjustment of operational holdings to specific assets. Households' resources, in particular their family labor endowment, influence the extent of land leased in. Our analysis revealed that households with relatively greater endowment of labor lease in relatively more land so as to ensure fuller utilization of family labor. In such a case, it is not surprising to find uniform labor intensity among tenure categories.

It has also often been argued that tenant households are more likely to depend on off-farm work than landed ones for two main reasons. First, the underdevelopment of the credit market is considered to be a crucial factor that prompts the tenant farmers to engage in the labor market. The flow of income during the year is dominated by the seasonality in agricultural activity. Seasonality produces income variability management problem that is aggravated when households lack financial intermediaries to provide lending and savings services. Due to their inability to offer the required collateral, it is difficult for tenant farmers to secure credit. Seasonal involvement of these households in off-farm activities is therefore an important way in which seasonal income is adjusted to the needs. Off-farm income can play a part as a substitute for farm credit that makes it possible to relax the capital constraint.

Second, tenancy contracts are motivated by the fact that the tenant is relatively well endowed with labor relative to land owned. When the land market function imperfectly, that is when farmers cannot fully adjust farm size to labor endowment through land rental or sales market, the resulting excess labor can be utilized efficiently through the labor market.

Table 2.7 confirm partly the differential occurrence of non-agricultural off-farm labor depending upon land tenure situation. The percentage of off-farm workdays appears as a decreasing function of the amount of the land owned only in Pitchit. In this province, landed households have increased their incomes by diversifying their agricultural activities whereas tenants have done so by engaging in the labor market. In Pitchit, the differences in the landholding sizes between land tenure categories are the highest, with tenants cultivating nearly half the area cultivated per worker by landed farms. It is thus likely that difficulties in access to land through the land market has limited the productive absorption of labor of tenanted farms into agricultural activities. In addition, tenants face a disadvantage in credit market compared to landed farms (Cf. Chapter 4). Hence tenants' off-farm activities, in Pitchit, reflect difficulties in access to both land and credit.

In Suphan Buri, the overall picture that emerge is one of remarkable similarity between the average characteristics of landlords and tenants after controlling for land cultivated. Members of landed households have a similar frequency to engage in off-farm and non-farm activities than tenant peasants. In this province, landholding sizes are almost equal among land tenure categories and very few tenanted farms appeared credit constrained. This would mean that the land and the credit markets function quite well in this province.

Table 2.7: Labor Input by Tenure

	Landowner/owner	Owner	Owner/tenant	Tenant
Suphan Buri				
Labor input on rice production	19.0	35.4	24.5	32.5
Family worker-days per rai	3.1	6.2	6.3	5.3
Hired worker-days per rai	17.3	16.6	22.1	16.8
Hired labor for rice in % of total labor				
Family labor allocation (%)				
Rice cultivation	58.3	64.5	66.2	73.3
Other agricultural activities	14.5	14.9	12.0	0.0
Non-farm activities	14.1	7.4	8.8	11.6
Off-farm labor	13.1	13.2	13.0	15.2
Pitchit				
Labor input on rice production				
Family worker-days per rai	19.4	25.7	15.9	37.2
Hired worker-days per rai	3.6	4.7	2.7	2.6
Hired labor for rice in % of total labor	17.6	21.5	14.5	11.9
Family labor allocation (%)				
Rice cultivation	65.9	70.0	72.9	81.3
Other agricultural activities	28.5	21.9	21.4	3.4
Non-farm activities	0.0	0.0	0.0	0.0
Off-farm labor	5.6	8.1	5.7	15.4

Roi Et				
Labor input on rice production				
Family worker-days per rai	61.4	57.2	30.1	50.4
Hired worker-days per rai	2.6	3.3	7.8	1.5
Hired labor for rice in % of total labor	8.8	11.6	23.2	12.1
Family labor allocation (%)				
Rice cultivation	100.0	78.3	78.5	80.7
Other agricultural activities	0.0	0.3	5.4	0.0
Non-farm activities	0.0	7.4	4.0	17.0
Off-farm labor	0.0	13.9	12.2	2.4

SOURCE: ORSTOM/CUSRI Project, Survey on Rice Farmers

In Roi Et non-farm self employment predominates over wage labor in the tenant category compared to other ones. This might reflect barriers to entry in the labor market for certain categories of workers. The distribution of workers by sector of activity and by land tenure shows that tenant farmers are agricultural laborers more often than any other class, which suggests that they have low opportunities of employment in the non-agricultural sectors. In the dry season, agricultural employment opportunities do not exist in Roi Et, whereas the scope for local market exchange tends to be low because rural industrial activities are limited. Supplementing farm income has to be done through the participation in outside labor markets. However, relatively poor and probably less educated households have a low propensity to acquire better jobs and encounter difficulties in diversifying their sources of income through salaried employment in the non-agricultural sector. The equilibrium position of these households whose workers lack the required characteristics to engage in the distant labor market requires therefore a greater reliance on home-based non-agricultural activities. Furthermore, the choice of occupation between agriculture and nonagricultural sector is generally associated with temporary change of residence which involves various costs besides the direct cost of travel, such as the cost of acquiring information about distant locations, changes in regional languages and culture, lack of extended family support in distant areas, and so forth. The costs of migration associated to a remote location might be unaffordable for the poorest.

2.4 Choice of Technology and Factor Proportions

2.4.1 The Seasonality of Agriculture and Seasonal Shifts in Labor Requirements

The requirements for labor input in rice cultivation are variable at different stages of cultivation, depending on the method of cultivation. Temporary work derives largely from the cyclical nature of rice farming activity and the related seasonal fluctuations in labor demand. Casual labor is generally expected during peak season requirements and is often regarded as a complement to family labor. There are basically four agricultural operations for which labor is hired: land preparation, seeding, harvesting and threshing[23].

Table 2.8 shows the relative importance of various farm operations in terms of the demand for casual labor that they generate. There are considerable differences in the operationwise distribution of hired labor input. Not surprisingly, demand for hired labor in rice growing is the highest in the transplanting and harvesting stages. Land preparation is not as important a task in terms of employment opportunity. Casual requirements for land preparation fall a long way behind transplanting and harvesting.

A comparison of hired labor inputs during the wet and dry seasons shows that there is apparently little difference in hired labor use per rai between the wet and dry season. The fact that the dry season transplanting and harvesting operation requires a slightly less absolute number of worker-days per farm than the wet season is mainly due to lower cultivated area.

There is a specific pattern of casual labor use in each province, depending on the method of cultivation (especially for planting) and on the overall productivity level of the farms. The study reveals that the most acute shortage of labor occurs in Roi Et during the transplanting period. Transplanting, which is the common practice in Roi Et, requires nearly 5 person-days per rai compared to 2-3 person-days where transplanting is mixed with broadcast sowing, as in Pitchit. In the Central Region a considerable part of the paddy fields are broadcast and there is little demand for wage workers for this operation.

---

[23] The labor component of crop care as well as irrigation and weeding has not been recorded and is thus not adequately reflected. Nevertheless, these tasks are generally performed by family labor rather than by seasonal workers.

Table 2.8: Hired Labor Input by Operation

	Suphan Buri		Pitchit		Roi Et	
	First crop	Second Crop	First crop	Second Crop	First crop	Second Crop
Number of worker-days per farm						
Plowing	7	7	8	11	16	
Sowing	6	6	95	57	121	
Harvesting	100	66	75	55	96	
Threshing	8	8	7	6	20	
Number of worker-days per rai						
Plowing	0.54	0.52	0.33	0.52	0.93	
Sowing	0.29	0.37	2.25	3.15	4.87	
Harvesting	3.39	3.10	2.70	2.19	4.15	
Threshing	0.37	0.40	0.25	0.38	0.63	

SOURCE: ORSTOM/CUSRI Project, Survey on Rice Farmers

A comparison of hired labor inputs during the wet and dry seasons shows that there is apparently little difference in hired labor use per rai between the wet and dry season. The fact that the dry season transplanting and harvesting operation requires a slightly less absolute number of worker-days per farm than the wet season is mainly due to lower cultivated area.

There is a specific pattern of casual labor use in each province, depending on the method of cultivation (especially for planting) and on the overall productivity level of the farms. The study reveals that the most acute shortage of labor occurs in Roi Et during the transplanting period. Transplanting, which is the common practice in Roi Et, requires nearly 5 person-days per rai compared to 2-3 person-days where transplanting is mixed with broadcast sowing, as in Pitchit. In the Central Region a considerable part of the paddy fields are broadcast and there is little demand for wage workers for this operation.

Difference in yields and the intensity of mechanization also explain the regional variations in hired labor use for harvesting operations. All the harvesting operations (cutting, assembling, carrying and threshing) consistently have the highest hired labor input per rai in Roi Et, though the yields are traditionally among the lowest in the country; capital investment in machinery has also been negligible. The Central Plain, which is the most progressive farming region with the highest percentage of irrigated area in the country, also has the highest level of farm mechanization. In this province, the impact of higher land productivity has been partly compensated by the introduction of machines. Finally, in Pitchit where both yields and level of mechanization are in a middle range, the hired labor use per rai during harvest time is the lowest.

2.4.2 The Impact of Technical Change on Labor Absorption

Labor utilization in rice farming depends on the production process, which is partly determined by ecological conditions and the rice-growing technology. Three types of technological factors can be distinguished: first labor-using factors such as high yielding varieties, application of fertilizer, which also require more careful water management and weeding. These technologies generally result in increases in output which, in turn, requires more labor in harvesting and post-harvesting operations. Second, land-augmenting capital factors such as irrigation which enhances labor absorptive capacity by increasing cropping intensity. Employment should thus expand more or less in step with cropping intensity. Third, labor-saving factors consist mainly of mechanization, which is said to substantially lower the need for labor, particularly in operations such as land preparation, harvesting, and threshing.

In order to assess the extent to which different types of technological packages affect labor absorption, meaningful production techniques were carefully defined. Five criteria were considered because of their significant effect on labor absorption: the number of crops per year (as a proxy for irrigation), the planting method, the use of fertilizer, the mechanization of land preparation, the mechanization of harvesting and post-harvesting operations. This resulted in twelve production techniques characterized by the different combinations of the criteria. The average labor input per rai associated with each production technique was then recorded in Table 2.9. The production techniques have been organized so that the intensity of labor use increases from the upper left of the table to the lower right.

The powerful impact of mechanization on labor inputs per rai is indisputable. Without machinery, harvesting is very labor-intensive. The family labor inputs per rai are dramatically cut as the level of mechanization increases, whatever the number of crops and the planting techniques. Non-mechanized rainfed and irrigated farms use twice as much family labor as their mechanized counterparts. The mechanization of rice farming operations also strongly affects the use of hired labor. There is a marked negative relationship between farm machinery use and the amount of hired labor for cultivation. For each planting technique, once the cropping intensity is constrained, mechanization results in a reduction in hired labor inputs. However, the amount of hired labor saved by mechanization varies according to the planting technique that the farmer employs. In the broadcasting group, the reduction is 53% for rainfed areas and 68% for double-cropped areas. In the transplanting group, the reduction is far less (14%), mainly because total pre-harvest labor use on non-mechanized farms does not differ much from that on mechanized farms since rice transplanters are not used in Thailand.

Table 2.9: Average Labor Input Per Rai for Different Production Techniques

	One crop/broadcast	One crop/transplant	Two crops/broadcast
No fertilizer/full mechanization			
Family labor input in rice production	9.9	12.5	
Hired labor input in rice production	0.7	4.0	
Number of farm days off	203.0	141.0	
Fertilizer/full mechanization			
Family labor input in rice production	15.4		28.6
Hired labor input in rice production	2.8		5.2
Number of farm days off	181.0		229.0
Fertilizer/mechanization of harvest			
Family labor input in rice production			51.1
Hired labor input in rice production			13.1
Number of farm days off			190.0
No fertilizer/mechanization of plowing			
Family labor input in rice production	27.5	62.5	
Hired labor input in rice production	3.8	2.8	
Number of farm days off	88.0	100.0	
Fertilizer/mechanization of plowing			
Family labor input in rice production	29.3	21.0	3.5.5

Hired labor input in rice production	3.9	5.4	5.5
Number of farm days off	194.0	135.0	69.0
Fertilizer/no mechanization			
Family labor input in rice production	45.0	54.8	
Hired labor input in rice production	4.8	3.4	
Number of farm days off	113.0	97.0	

Source: ORSTOM/CUSRI Project, Survey on Rice Farmers

Land-augmenting technologies have strongly influenced the demand for both family and hired labor. Employment is significantly influenced by the effective area under cultivation. Double-cropping tends to raise labor inputs depending on the level of mechanization. Irrigated non-mechanized farms used 80% to 120% more family labor than rainfed non-mechanized operators, while irrigated non-mechanized holdings used only 20% to 40% more of this labor than their mechanized counterparts. The rise in labor use as a result of increased cropping intensity is thus partially mitigated by the mechanization of cultural operations. Likewise, cropping intensity is found in the data to correlate positively with casual hired labor. Farmers in non-irrigated areas who grow rice in the rainy season only use far less hired labor per rai than farmers who grow rice in both the rainy and dry seasons.

The differences in labor absorption related to changes in the planting technique are also striking: transplanting is usually associated with a rise of 10% to 25% in family labor use. Interestingly, there is a much more marked difference in use of hired labor input between transplanting and broadcasting with broadcast rice requiring, on average, half of the hired labor used for transplanting. Finally, use of fertilizer tends to increase labor requirements. A farm that applies fertilizer requires an average of 30% more casual labor than farms that do not.

Finally, the amount of labor farm families can supply to the labor market should be influenced both by the ecological setting and by the technology used for rice farming, which in turn determine whether such employment competes with farm work. Farmers who grow rice once a year due to lack of water have theoretically more time to spend in off-farm activities compared to those who grow rice twice a year. Data in Table 2.9 show that a more intensive cropping pattern does not seem to discourage off-farm work. On average, households do not report lower off-farm employment activities when land-intensive techniques are carried out. Mechanization should also push workers released by machines to look for jobs outside the farm. This appears to be the case in our sample of rice farmers. The use of agricultural equipment is associated with heavier involvement of family labor in off-farm economic activities.

To summarize, higher cropping intensities are associated with higher labor employment throughout the year, both on and off the farm. This brings about a more stable seasonal pattern of employment, which, of course, has positive welfare implications. By contrast, a negative association can be observed between on-farm labor-intensity and capital intensity, high capital intensity being accompanied by low family and hired labor use. In highly capitalized farms, out migration of workers in search of jobs solved the problem of mismatch between labor force availability and employment requirements.

2.5 Interactions between Rural and Urban Labor Markets

2.5.1 An Overview of Off-Farm Activities

Since Lewis's article on surplus labor, considerable research has been generated on the question of a dual labor market in developing countries. According to Lewis, an unlimited supply of labor may be said to exist where population is large relative to capital and natural resources, so that there are large sectors of the economy where the marginal productivity of labor is negligible or even equal to zero, leading to the existence of « *disguised* » unemployment. A prominent feature of labor market dualism is thus high unemployment and/or unemployment in the rural economy.

This hypothesis has already been documented in Thailand. The first studies examined the seasonal unemployment and underemployment issue over the last several years and rejected the notion of a huge reservoir of underemployed rural workers, although there may be some regional variations (Bertrand and Squire, 1980; World Bank, 1983). Instead, they concluded that the labor market in Thailand works well and labor mobility in response to differential employment and earning opportunities is high. More recent studies, however, (Sussangkarn, 1987; Sussangkarn, 1990; Sussangkarn and Chalamwong, 1994), contradicted these previous views and showed that seasonal unemployment is a problem that afflicts a great number of people. Most of the seasonal unemployed, especially younger workers and females cannot be regarded as voluntarily withdrawing from the labor force.

The general pattern, which emerges from our data, is that of a high level of labor force participation and a large number of days worked throughout the year. Of course, there is a significant change in labor utilization patterns among rice farmers between the peak and slack agricultural season, with the dry season agricultural labor force contracting compared to the rainy season labor force. However, open unemployment has been found to be extremely low both in double cropped and rainfed areas; in the former, family laborers work on the farm while in the latter a shift to off-farm employment by members of farm families is observed during the slack agricultural season. Off-farm work is widespread among rice farmers with nearly one half of farm households having at least one member who hold some salaried job outside the holding.

Household member's occupations in off-farm jobs are shown in Table 2.10. Of the occupations that family laborers are concurrently engaged in, the most common is agriculture in Suphan Buri and Pitchit, which absorbs 45% to 60% of off-farm workers. In Roi Et, less than 10% of the households' members had agriculture as an off-farm activity. Construction is the second most common occupation in Suphan Buri and Pitchit with 20-23% of household members being engaged in this occupation. Manufacturing and transport are found to provide most of the jobs to the household members of Roi Et who are engaged in off-farm work; factory workers, carpenters, tailors, dress making, silk and cotton weaving and taxi drivers are in this category. Involvement of outworkers in the service and commerce sector is minimal in all provinces. Many of the household members engaged in commerce are small merchants dealing primarily in agricultural produce, farming materials and other daily necessities.

Table 2.10: Percent Distribution of Off-Farm Workers by Industry

	Percent of Distribution of Workers			Percent of Distribution of Workers		
	Suphan Buri	Pitchit	Roi Et	Suphan Buri	Pitchit	Roi Et
Agriculture	46.4	59.1	9.3	27.8	36.8	7.0
Construction	20.3	22.7	13.0	8.4	22.9	15.2
Manufacturing	14.5	4.5	27.8	25.4	11.9	21.3
Transport	7.2	3.0	31.5	12.9	6.7	28.0
Commerce	2.9	9.1	7.4	5.7	16.1	9.5
Services	8.7	1.5	11.1	19.7	5.7	19.0

SOURCE: ORSTOM/CUSRI Project, Survey on Rice Farmers

The prevalence of agricultural off-farm activities in Suphan Buri and Pitchit requires some comments. Non-agricultural job opportunities are expected to be relatively greater in better developed villages due to availability of infrastructure facilities and greater demand for non-agricultural goods and services arising from higher levels of urbanization, industrialization and incomes. Suphan Buri is closer to Bangkok than Pitchit and Roi Et, its villages are usually close to towns; off-farm employment opportunities in the non-agricultural sector are therefore quite great and easy to get to. One would therefore expect the proportion of family members engaged in non-agricultural activities to be higher in Suphan Buri than in Pitchit and higher in Pitchit than in Roi Et. Actually, the more industrialized provinces have the highest proportion of workers engaged in agricultural off-farm activities. The question that arises is thus why many outworkers engage in agricultural work in Suphan Buri and Pitchit while in Roi Et, only relatively few do so?

One rationale is that in Suphan Buri and in Pitchit, where investment in irrigation has been concentrated, a large part of off-farm activities are related to the strong demand of farm households for agricultural labor both in the wet and dry seasons. In contrast, Roi Et has the lowest percentage of workers engaged in off-farm agricultural jobs because there is no labor demand in the dry season due to the lack of water; family members cannot find substantial amounts of market work as hired agricultural laborers. Therefore, most rural households in this area can participate in off-farm activities during the slack season only by shifting economic activity and migrating to the cities. Another explanation discussed in more details later is that the wage differential between agricultural and non-agricultural activities, especially during the second crop cycle, is far less in the double cropped areas than in rainfed areas.

Another striking feature of rice farmers' off-farm activities, according to their location, is the absence of concentration in Metropolitan Bangkok. It is obvious from Table 2.11 that the bulk of temporary migration into a district originates from the immediately neighboring districts. In Suphan Buri and Pitchit, nearly 50-70% of family workers move within the same district and another 10-25% within the same province to find temporary jobs. As might be expected it is Roi Et, the province the most dependent on rainfed agriculture, with the lowest income levels and the fewest alternative employment opportunities in regional urban centers, where the seasonal outflow of labor to Bangkok is the greatest. In this area, the nature of rural manufacturing which makes up the bulk of non-agricultural employment is quite specific and limited (Biggs 1990; World Bank 1983). Small manufacturing establishments are therefore a limited source of income and employment for northeastern rice farmers. Opportunities for greater participation of the rural labor force in off-farm local industrial activities will probably have to wait for more manufacturing growth in cities. Migration from arid Northeast to Central urban areas therefore makes sense, and is happening on a substantial scale.

Table 2.11: Percent Distribution of Off-Farm Workers by Location of the Job

	Percent of Distribution of Workers		
	Suphan Buri	Pitchit	Roi Et
Same district	55.7	67.9	11.3
Same province	25.7	7.1	12.9
Another province	11.4	7.1	17.7
Bangkok	7.1	17.9	58.1

SOURCE: ORSTOM/CUSRI Project, Survey on Rice Farmers

A conventional view about off-farm activities is that they are carried out by farmers during times in the seasonal cycle when labor is in surplus. Typically, it is during the slack season, when cultivator households have less work on their own farms, that they are more willing to engage in off-farm labor. Table 2.12 indicates that nearly 40% of family members who took jobs outside the farm worked off their holdings for three months or less, another 40% did so for 3 to 6 months, 20% only for more than 6 months. Some activities are obviously conducted in sequence while others are conducted alongside farming activity.

Agricultural laborers tend to be employed on short-term contracts while participation in non-agricultural activities is usually much longer. The longer jobs usually refer to governmental officials, officials of agricultural cooperatives and even school teachers. Comparing the three provinces, there is a larger percentage of part-time farm households with stable concurrent jobs in Suphan Buri. The sequential system is found in zones with one rainfall season per year.

Table 2.12: Percent Distribution of Off-Farm Workers According to the Length of the Job

	Suphan Buri	Pitchit	Roi Et
Less than 3 months	34.2	56.0	14.5
Between 3 and 6 months	34.2	35.7	67.7
More than 6 months	31.5	8.3	17.7

SOURCE: ORSTOM/CUSRI Project, Survey on Rice Farmers

Finally, it must be emphasized that information on employment opportunities is generally transmitted through personal contacts. Search procedures for the most part rely on relatives, friends and acquaintances. Kinship networks exist which tend to lower the risks associated with migration by providing important information to would-be migrants about labor market conditions. On the employer's side, a popular means of recruiting is to ask current workers to bring their friends and relatives. Hence, information and opportunity channels often emerge because of prior out-migration from certain geographic communities, increasing the likelihood that other migrants will follow. As a result, a fairly sizable proportion of migrants has prearranged jobs and arrives at the place of work with an informal promise of a job. Most striking in the job search process is the unimportance of employment agencies and their irrelevance in the migratory process, a fact confirmed by other studies (Phananiramai, 1994). Government labor offices are of limited use as well.

Table 2.13: Distribution of Workers According to the Job Search Process

	Suphan Buri	Pitchit	Roi Et
Worker's initiative	26.5	18.1	36.2
Friends	32.4	20.5	20.7
Relatives	8.8	16.9	24.1
Employment Agencies	2.9	6.0	-
Middlemen	11.8	6.0	15.5
Employer	17.6	32.5	3.4

SOURCE: ORSTOM/CUSRI Project, Survey on Rice Farmers

2.5.2 Labor Market Adjustments and Wage Rates

2.5.2.1 Measuring Agricultural Wages

The analysis of the wage differentials employs two kinds of data from the survey. First, there are observations on the costs incurred by the farmer for operations performed on each farm during each crop cycle. Second, there are observations on the daily wage paid to outworkers throughout the duration of the job.

As far as data on labor costs are concerned, several adjustments were made in order to form the basis of an evaluation of the daily agricultural wage. A first problem we had to deal with arose from the existence of a broad range of mechanisms for recruiting extra-familial labor for rural work beyond a clear wage contract: there are communal work groups operating entirely on the basis of generalized reciprocity and others in which some work only for food; lending of plow or gifts of food at harvest time in return for occasional labor is sometimes observed. In most cases, agricultural workers are paid partly in cash, partly in kind in the form of meals, a homestead to live in provided by the employer, or as a share in output. The form of payment is usually related to the spatial proximity to an urban labor market, the nearer to the market the greater the cash component of the total earning. Consequently, the total payment recorded in the survey has been differentiated with respect to wage and non-wage benefits. Kind components were converted into cash equivalents.

There are basically two types of labor contracts in Thailand: piece rate and time rate. Remuneration based on piece rate contracts is the most widely used in each province surveyed. Given the large range of units used for the different operations[24], daily wages had to be derived using data on yields, the area cultivated and a rough work intensity coefficient measured in full and half day units. The use of this coefficient raised another problem coming from the fact that in peak seasons, workers sometimes work for longer than a usual full day. In this case, the earnings show up in our data as higher wages, whereas the actual wage per standardized day remains the same.

---

[24] The costs of plowing and cutting were generally given per rai, the cost of assembling per sheaf, the cost of carrying per trip, the cost of threshing per ton, etc.

Then, it can be argued that part of the difference in nominal wages is illusive, because of the different cost of living in different places. As for regional differences in cost of living, the Bank of Thailand publishes index prices for each region. We found that the differences across regions are very small: the price index in the highest-cost region (Central Plain) is only 3% above that in the lowest-cost region (North). So the use of deflated wages on the grounds of spatial variations widens the regional income disparity only slightly more than the use of nominal wages. Adjusting for a lower cost of living in rural areas compared to urban ones would therefore leave substantial average wage differentials among unskilled workers.

As far as the daily wage rate is concerned, one should be careful when comparing the figures between industries because for self-employed entrepreneurs, which are found mainly in the commerce and transport industries, average daily earnings might reflect returns on factors other than labor.

Lastly, it is important to keep in mind that the wide variations along the spatial, seasonal, contractual and quality dimensions of employment inevitably produce a correspondingly wide dispersion of wages and reduces the significance attached to measures of average wage levels. Variations in remuneration is economically and socially embedded in age, gender, and human capital differences among laborers. Accordingly, the bargaining strength of different persons will vary, with the result that the price paid for the same factor of production will differ markedly from one person to another. In addition, different operations, timing of hiring, different duration of employment and forms of labor contracts lead to different wages. Finally, individuals are not homogenous in their provision of effort. It is obvious that individuals who provide different levels of effort will receive different wages. By and large, there is no such thing as « *the* » price of a factor of production in rural areas: different groups of people often face radically different sets of relative factor prices.

2.5.2.2 Regional and Operation-Specific Agricultural Wage Differences

Table 2.14 presents the estimation of the daily wages received by hired laborers and calculated according to the costs of cultivation reported by employers. It gives some idea of the magnitude of seasonal earnings received by workers in specific operations over the crop cycle in paddy cultivation. Contrary to the implications of the popular theories on wage determination by biological factors (subsistence theory wages), the agricultural wage in the rice sector appears sensitive to demand and productivity conditions. There is ample evidence from field observation that wages for farm labor do vary significantly within localities from season to season and within seasons from one locality to another. There is on the whole a positive association between the wage rate and demand expansion factors, both through agricultural growth and across seasons.

Table 2.14: Variations in the Daily Wage Rates for Selected Agricultural Operations

	Suphan Buri	Pitchit		Roi Et
	Irrigated Rice	Irrigated Rice	Rain fed Rice	Rain fed Rice
First crop				
Land preparation	98	37	36	30
Planting	94	57	60	43
Cutting	106	102	87	46
Carrying/Assembling	105	90		88
Threshing	150	142	132	78
Second crop				
Land preparation	90	62		
Planting	111	78		
Cutting	113	108		
Carrying/Assembling	131	123		
Threshing	158	181		

SOURCE: ORSTOM/CUSRI Project, Survey on Rice Farmers

The rice labor market in Thailand appears particularly tight during the peak agricultural season, which occurs once or twice a year, depending on the cropping intensity. For instance, during the harvesting season, when many farmers are busy with their own harvest, casual wages are double what they are in the period of land preparation in Pitchit and Roi Et, and increase by 25% in Suphan Buri.

Interestingly, there is an indication of significant differentials in wage rates between first and second crops, with second crop wages significantly higher than first crop wages. That agricultural and non-agricultural sectors compete for scarce labor at that time explains this scenario. Demand factors in the non-agricultural sector tend to draw laborers from rainfed farming areas into urban jobs. Labor for rice cultivation is therefore in relatively scarce supply.

Substantial cross-province differences exist in the daily wages of agricultural workers. Higher agricultural productivity is associated with higher wages for agricultural workers. Technological changes such as chemical biological innovations or higher cropping intensities and the resulting increases in yields have raised the productivity of farm labor, which in turn has translated into higher wages. The wages in Suphan Buri appear to be higher than in any other province. In irrigated areas of Pitchit, wages have responded only to the increased demand for labor during the harvesting period, whereas wages for land preparation and planting have not changed relative to those of rainfed areas. In the Northeast, recurrent drought and persistent poverty have served to cheapen labor and sometimes farms are able to recruit workers without paying more than 30 baht per day.

A frequent assumption on wage disparities is related to differences in the monopsonistic or oligoponistic power the employer is supposed to exert in fixing the terms of the labor contract. Some authors (Bardhan, 1979) expect that the higher the degree of monopoly in the land market and the more severe local unemployment is, the lower the alternative opportunities the workers face, the lower the wage will be. This will occur for two reasons. First, the greater land concentration, the more likely it is that landowners will cease to be price takers in the labor market. Second, the greater the land concentration, the more labor is likely to be on offer and the presence of a large supply of labor results in a lowering of wages. Considering the same argument from the demand side, other authors (Rosenzweig, 1978) argued that the demand for wage labor should be the highest in areas where landholding inequality is greatest. Where the distribution of land holdings is more unequal, more land is held by labor-importing farm households, which in turn tends to raise wages.

As far as labor for rice cultivation is concerned, inter-province mobility is sufficiently low, so that province-level characteristics are important determinants of the province wages and market labor supply. In light of our data, rural wages and the measure of both landholding and landownership inequality, through the Gini coefficient, are positively associated. Where landholding inequality is greater, wages are higher. This interesting result, which confirm Rosezweig's view, is also explained by the fact that even where the employers could dominate the village market for labor hire, potential laborers still have access to some alternative uses of their time in non-farm occupations in their villages or in migration to other villages or nearby towns. Hence, even if local employers may have some monopsony power, this power is not enough to constrain wages to low levels.

Finally, the unavailability of data on workers' acquired attributes such as education and experience did not allow us to investigate empirically whether these variables have any impact on wage differentials. However, considerations with respect to human capital might not be very relevant in explaining wage differences since the rural wage workforce consists mostly of homogeneous group of unskilled workers. Literacy or education is not considered an important factor in hiring male or female wage workers for operations like planting, weeding, harvesting and so forth, especially when most operations are labor-intensive and performed manually.

2.5.2.3 Agricultural/Non-Agricultural Wage Differences

There is often a perception that wages in rural areas are massively lower than in urban centers so that that rural labor is pulled into non-agricultural activities by wage opportunities, which are considerably above the supply price of labor from the rural sector. In order to provide some idea of the returns to wage employment in the rice sector and in urban industries, data on daily wages received by family workers are presented in Table 2.15 according to the branch of industry and the location of the job. Agricultural as well as non-agricultural wages here are those reported by family laborers who where engaged in temporary off-farm activities in each province, as well as in Bangkok.

In general, although the non-agricultural wage rates were found to be higher than the wage rates in agriculture, the gap is variable across provinces. The data suggest that rural labor wages in Suphan Buri are higher than commonly assumed, especially when compared to non-agricultural wages in local cities. There is little wage differential between unskilled urban workers and agricultural workers in this Central Plain province. Conversely, urban wages in Pitchit and Roi Et are nearly double rural wages.

Table 2.15: Agricultural and Non-Agricultural Wages

	Suphan Buri	Pitchit	Roi Et	Bangkok
Males				
Agriculture	92	51	45	
Manufacturing	106	100		126
Construction	119	84	60	112
Commerce	100			200
Transport	189	100	55	200
Services	200		140	120
Females				
Agriculture	78	53	45	
Manufacturing	83			120
Construction	100			
Commerce		143		130
Transport		74		
Services			118	

SOURCE: ORSTOM/CUSRI Project, Survey on Rice Farmers

Inter-province wage differences are quite significant and the largest gradient is between the outer regions (North, Northeast) and the Central Plain. Agricultural wages in Suphan Buri are roughly 1.5 times the level of Pitchit, whereas rural wages in Pitchit are 15% higher than in Roi Et. The differences among provinces in non-agricultural wages seem as pronounced as those for the agricultural wages. For example, wages paid in the construction sector in the North and Northeast are roughly 30 to 50% below those paid for the same type of labor in the more developed Central Plain. The Central Plain thus appears as a higher-wage area for both rural and urban labor.

Wage rates in Bangkok appear, by and large, higher compared to those prevailing in the provinces. However, the data indicate little difference in the wage rates in Bangkok and the cities in the Central Plain. This explains why rural laborers in Suphan Buri and to a less extent in Pitchit do not migrate over long distances and why migrants from the Northeast represent 65% of workers migrating to Bangkok in the sample. Rural laborers from Suphan Buri and Pitchit were found to earn higher wages in local jobs than did their counterparts from the Northeast, whatever the branch of industry.

Table 2.15 also presents some intriguing characteristics of the labor market in relation to the gender wage earnings gap. Suphan Buri with relatively high level of yields and agricultural development is characterized by a higher male-female wage differential than are the other two poorer provinces where there is very little or no significant differential between men and women engaged in farm labor. Female agricultural wage workers earn only 85% of male workers' wages in Suphan Buri, whereas in Pitchit and Roi Et the differential appears non-existent.

It has often been said that in some agricultural operations a wage premium for muscle power is expected, and/or that male laborers are more productive than female ones. There is, however, no evidence to suggest that female farm wage workers are less productive than male workers or that farm women cannot or do not perform as hard physical labor as men do. It is more likely that the introduction of mechanical technology has influenced the productivity of male workers more than female workers, because most farm machines are operated by male workers[25]. As a result, wage differentials between men and women have widened in more mechanized areas of Suphan Buri, whereas in other areas where mechanization is still in its early stage, a wage gap between male and female wage workers is not observed.

As far as non-agricultural jobs are concerned, the data indicate that there is a significant wage differential by sex of some 20 baht a day in almost all sectors of activity, a phenomenon that holds across the three provinces studied and for workers who have temporarily migrated to Bangkok. The traditional explanation for the lower expected wages of women is that women are encumbered in their supply of labor services by various social and economic restraints related to the household economy, which make them irregular suppliers of labor. Off-farm work means prolonged absence of women and families are deprived of the benefits of women's domestic work. For women, wage employment may not be easy to obtain or may be available at inappropriate times and/or for inappropriate duration, resulting in limited choice of off-farm activities.

[25] Another reason might be linked to the fact that men sometimes provide the machines they drive. Thus, in this case, the earnings represent returns on both machine and human labor power.

The pattern of off-farm activities and migration observed in the sample is consistent with the pattern of wages received. Residents of different areas supply different levels of off-farm labor because their benefits from off-farm work differ. High urban wages in the destination place have a positive impact on the rural-urban migration, whereas high rural and urban wages in the province of origin tend to reduce the flow of out-migration. The large flow of migrants from Roi Et to Bangkok can be attributed to the large gap between agricultural wages in that province and the returns on migration to Bangkok. Conversely, in Suphan Buri, where agricultural and non-agricultural wages are high, most family workers found occasional jobs in the province.

2.6 Conclusion

Rice farmers in Thailand face a wide range of choice regarding their participation in economic activities: they can use more labor productively in rice farming by intensifying methods of cultivation; they can also construct a mixed portfolio of activities including diversification of agricultural activities, development of non-farm activities, and increase in labor supply into off-farm activities. A delicate balance therefore occurs in the use of family labor resources among farm, non-farm and off-farm activities as well as in the source of labor for rice farming (family and/or casual). Most households are extensively involved in buying and selling labor and distinctions between laborers and employers are irrelevant. Household members often switch status from one season to the next. However, net employers typically own and cultivate more land whereas net laborers cultivate small rented areas.

The data collected from the field survey highlight the general trend for per hectare labor input to decline as holding size increases. This means that the Thai rice sector, made up of typically small family farms, is structurally labor-using. The results here tend to suggest that alleged imperfections in the labor market in Thailand are not mainly responsible for the negative relationship between farm size and labor absorption. Exogenous differences in land quality as well as disparities in the cost of capital borne by households of different size, led to different combinations of input and output ratios.

The choice of technique has a considerable effect on the amount of labor used during the year. Twelve different combinations of planting techniques, water management and sources of power have been examined. Overall labor use per rai on mechanized farms has been found substantially lower than on non-mechanized ones. In contrast, the switch from one crop to two crops as well as the choice of transplanting compared to broadcasting, results in sizeable increase in employment, whatever the source of power.

On the whole, the technological changes observed in rice farming did not contributed much to restrain the demand for labor in aggregate terms. Mechanization, whose impact *per se* would result in a high loss in employment opportunities, has been accessible only to some farmers, especially those in the Central Region and adjacent areas. In addition, mechanization has been associated with double cropping so that the labor displaced by machines has been reabsorbed in the cultivation of the second crop. As a result, the shifts from "traditional" to more "modern" techniques that took place during the past decades in irrigated areas entailed net increases in labor requirements.

Technical change has been correlative to rising levels of real wages in the provinces concerned. Our results confirm the expected positive association of the wage rate for farm labor with demand factors such as irrigation intensity, the level of agricultural development and the period of the year. However, agricultural growth has not been the only cause of increased real wages. The tightening of rural labor markets has also been brought about by increases in non-agricultural productive activities. The development of urban centers offering alternative employment opportunities for agricultural labor had important repercussions on the demand for labor and the trend towards the decline in the agricultural labor force has been associated to an upward adjustment of agricultural wages.

The extent of involvement of the rice households in off-farm activities and the time allocation decisions between farm and off-farm activities have been discussed. The evidence from the field survey suggests that both «*pull*» and «*push*» factors have been at work. The seasonality of agricultural labor requirements, a large family size, a sex ratio biased toward males, a large number of dependents per working adult, a small landholding, have been the main «*push*» factors. As far as "pull" factors are considered, the involvement of rice farm households in off-farm activities has been found positively associated with wage differentials but quite independent of the conditions in nearby urban market centers, and of the distance from the village to urban centers. The development of an improved transportation system and the expansion of the means of communication have obviously facilitated the responsiveness of workers to wage differentials between different places.

Finally, it is worth underlining the importance of agriculture as a source of off-farm activities in certain areas. This suggests that policies that detract from agricultural prosperity may have a powerful influence on off-farm rural employment and incomes.

Chapter 3

CAPITAL FORMATION

3.0 INTRODUCTION

Technology adoption by agricultural producers is viewed as an essential prerequisite for economic prosperity in less developed countries because agriculture is an important source of these countries' gross domestic product and constitutes a major share of their exports earnings. Indeed, increasing productivity through technology adoption is a primary mechanism for sustaining longer run profitability in agriculture because technological change offers a unique opportunity to overcome the constraints imposed on agriculture by natural conditions and limited supplies of scarce factors.

There are basically two kinds of capital investment, land-augmenting and labor-displacing investments. Irrigation combined with the predominant use of improved seed varieties and the more concentrated use of other farm inputs, e.g. fertilizer and pesticides, which contribute to generally higher yields are examples of land-augmenting technologies. Investment in machines like tractors is useful mainly in replacing labor.

In theories of endogenous technical progress, population growth and population pressure on land have been seen as important sources of technological change in agriculture (Boserup, 1965). According to Boserup, as population density increases, populations cannot be fed any more without capital investment in land improvement. Changes in land use from shifting cultivation to annual cropping and later multiple cropping are thought to be driven by population growth. The model of induced innovation suggested by Hayami and Ruttan (1985) and Binswanger and Ruttan (1978) provides an economic theory of innovation whereby relative price changes determine responses by economic agents and research institutions. Given a technology consisting of a set of techniques with various factor-proportions, the technique which economizes on the use of factors which have become scarce, as indicated by relatively high prices and intensifies the use of other factors whose prices are relatively low, stands every chance of being adopted.

The relatively rapid growth in Thai rice output in the past decades has been achieved by adding more labor and land to production; productivity increases have been of minor importance (Phélinas, 1997; Islanonda and Poapongsakorn, 1995). However, the scope for an increase in rice production by bringing more area under cultivation is limited because the land frontier reached in the early 1980s. Meanwhile, Thailand has experienced a declining rice farming population due to both a contraction of population growth and an impressive increase in nonagricultural employment opportunities. As a result, Thailand's traditional economic advantage in abundant land and cheap labor has been significantly eroded.

The form and intensity of the response to this progressive factor scarcity at the local level is the subject of this Chapter. Through a statistical analysis of the relationships between differences in relative factor use and relative factor prices, we try to interpret the process of technical change in Thai rice farming. We describe the patterns of adoption of modern agricultural capital and give an overview of the key features of rice farming technology in Thailand, covering the degree of mechanization, the use of modern inputs, the extent of irrigation. The factors affecting the adoption of technical innovation are examined. Since the determinants of investment in labor-saving machines are expected to be different from those influencing the purchase of land-saving current inputs, we make a sharp distinction between these two types of capital goods.

3.1 The Failure of the Green Revolution

Whereas the years from the mid-1960s to the present have been a period of particularly rapid change in the agriculture of South East Asian countries, the results that occurred in Thailand contrast sharply. The transformation of rice production technology induced by the new high-yielding varieties had only a minimal impact on rice cultivation in Thailand. A large share of the rice producers have been left untouched by this technology, particularly those dependent on rainfall or affected by problems of soils and lack of control over water supplies.

The best indicator of the intensification of production is the level and rate of change in paddy yields. Paddy yields in Thailand have not improved significantly; they showed a downward trend from the beginning of the century until the Second World War and have then increased by less than 1% per year over the past four decades. Yields of wet season paddy remained roughly constant (0.3% per year) from 1960 to 1996. Changes in yields were more dramatic in dry season farming; they grew by nearly 2.5% per year over this period.

One of the rationales for the stagnating trend of wet season paddy yields is that since the Second World War new and often less fertile lands have been brought under cultivation. The increasing share of regions outside the Central Plain in the area under paddy during the 1960s and 1970s accounted for a large part of the national stagnation in wet season yields. In contrast, most of the increase in the dry season yields can be attributed to the rapid yield increase in Central Plain, where the provision of irrigation has enabled increased use of fertilizer and other modern inputs.

There can be little doubt that the paddy yields depend on environmental factors such as soil quality, rainfall and drainage. Indeed, environmental factors impose a series of threshold levels below which any intensification of agricultural production cannot take place, or at least severely limit the economic viability of intensification. Nevertheless, the bio-physical conditions which characterize agriculture in any given region are not invariant over time, nor determined solely by factors beyond any kind of human control. Thus, while natural bio-physical conditions may limit the production frontier in the short run, these conditions can themselves be altered in the longer run.

These considerations suggest that the lag in intensification in Thailand is not due to special environmental conditions. Some neighboring countries (Burma, Malaysia, Vietnam) face climate conditions like Thailand's yet their average yield level is much higher. Poor performance in Thai paddy yields can also be attributed to the lack of public investment in irrigation, strong taste preferences of Thai consumers for low yielding traditional rice varieties, and decades of taxation on rice which have inevitably kept the intensity of input use at a low level.

3.1.1 Irrigation and Water Management

Irrigation serves as a factor increasing the effective supply of land: a piece of irrigated land may be regarded as a multiple of a piece of unirrigated land of the same acreage. But above all, the development of irrigation facilities is considered as a pre-condition for adoption of more intensive techniques, mainly because irrigation reduces production uncertainty. For the farmer, the risk of crop failure is not a matter merely of decline in output; it is one of losing the inputs invested and of decline in income.

Since the potential for changes in production techniques is greatly dependent on improved water control, we begin the discussion with the existing state of the irrigation infrastructure. Irrigation investment has taken two forms in Thailand: the first one consisted of major engineering projects usually carried out by governments on the basis of their perception of social costs and benefits. The other one consisted in the installation of small mechanical pumps, which are private investments in nature, and were undertaken by individuals in response to their perceptions of private profitability. Both private and public sector investment in water control facilities are examined here.

3.1.1.1 Thailand's Environmental Conditions

The productivity effects of irrigation are quite different depending on climatic conditions and on the effectiveness of the particular types of irrigation facilities concerned. The amount of irrigation needed to bring a crop to harvest depends on the amount and monthly distribution of local rainfall, soil moisture reserves, soil type and potential evapotranspiration, which is the maximum amount of water that can be lost through surface evaporation and transpiration. An index of irrigation requirements can be derived by comparing rainfall with evapotranspiration. The greater the shortfall in rainfall in relation to evapotranspiration, the greater the irrigation requirement.

Thailand's territory fall into climatic conditions that are, in general, marginal for rainfed rice cultivation (Ishii, 1978). The annual monsoon defines a rainy season that lasts from May through October and although rainfall is quite abundant, its distribution is uneven in both space and time. Even in the Central Region, which is held to have excess water, the water supply is extremely unstable and as far as paddy cultivation is concerned, tends to be insufficient (DORAS, 1996). In all the other regions average rainfall falls short of average crop water requirements in most months of the year and the effects of drought can be severe. Hence, without water control, only one crop a year is climatically possible in most rice growing areas of Thailand. In addition, the instability of the climate and water conditions adversely affect the wet season rice growing in areas where suitable irrigation and drainage facilities are not available: shortage of rainfall in April and May may delay planting, the distribution of water from May to August is highly unpredictable and may prevent normal growth, and finally, flooding following occasional heavy rains may damage the crop. Supplementary irrigation is thus a necessity for adequate wet season harvests even in normal years.

3.1.1.2 Thai Irrigation Policy

Historically, the Thai government has been the main source of finance for large and medium-scale irrigation and water conservation projects. There has been relatively little private investment in hydraulic infrastructure. The desirability of irrigation has long been recognized by public authorities in Thailand. Until the late nineteenth century, the use of *corvée* labor as well as hired Chinese labor served to built canals for both the distribution of flood waters and transportation. The main drawback of this system was that it was not able to deliver water over the amount made available by the flood. The beginning of irrigation, as opposed to distribution of flood waters dates from the well-known Van der Heide mission, engaged in the early twentieth century to design an overall irrigation plan for the nation (Ingram, 1971; Feeny, 1982). Unfortunately, the ambitious and costly Van der Heide proposals made in 1902 and the reduced project proposed in 1906 were both rejected by the government. The decision of the government to invest funds in railways rather than in irrigation was the result of a conflict between national security and economic development goals, (Feeny, 1982). Railroads provided strategic benefits and served administrative goals mainly because they allowed the government to enhance Bangkok's power over provincial and local government. Consequently, even though irrigation was an attractive public investment, few works were carried out compared with the huge potential of Thailand (6 million hectares) until the end of the World War II; they were mainly located in Central Plain which benefited nearly 650,000 hectares.

Following an FAO mission in 1948, the government finally started to invest in increasing agricultural productivity. The Van der Heide project (under the new name of Greater Chao Praya Project) was finally implemented with World Bank assistance. Irrigated area increased by nearly 3% per year from 1960 to mid-1975, and then irrigation area expanded at an annual rate ranging between 3% and 5% up to the mid-1980s (cf. Table 3.1). Since the mid-1980s, the focus of irrigation development has been shifted from a strategy that emphasized constructing large-scale irrigation systems to one that helps existing systems to improve their performance. Therefore, the share of the construction budget in the annual budgets of the Royal Irrigation Department has declined (TDRI, 1992), and the growth of irrigated area has slowed considerably to 1.7-1.0% per year.

Table 3.1: Trends in Irrigated Area, Rice Labor Force, and World Price of Rice

	Annual growth rates			Average real
	Irrigated area	Rice labor force	Cultivated land per rice worker	world price of rice ($/ton)
1920-1955		2.07	0.34	
1955-1960		2.59	0.78	
1960-1965	2.0	2.24	-1.60	
1965-1970	2.9		-0.50	554
1970-1975	3.2	2.76	-0.62	634
1975-1980	5.4		-1.17	579
1980-1985	4.2	1.84	-1.30	388
1985-1990	1.7	1.16	0.00	259
1990-1995	1.8	0.00		258
1995-1999	1.0			192

SOURCES: Office of Agricultural Economics; Feeney, 1982; Labor force Surveys

Estimates of irrigated area for rice show that no more than a quarter of the paddy land receives some irrigation benefits (Isvilanondan and Poapongsakorn, 1995). In addition, irrigation network is largely concentrated in the Central Region, a reflection of the topographical advantages of the Central Plain, and is relatively insignificant in the North and Northeast. The irrigated area covers 44% of the cultivated area in the Central Region, while the corresponding figures are only 10% in the North and less than 1% in the Northeast.

3.1.1.3 An Analysis of Irrigation Patterns and Trends

The presence of population-driven technological progress in the agricultural sector has been extensively discussed. Boserup (1965) has provided ample evidence to support a model of endogenous technical progress in agriculture that predicts intensification in response to increased population pressure. There should thus exist a close relationship between population growth and investment in irrigation. Table 3.1 presents estimates of the labor force engaged in rice farming, irrigated area, and the ratio between the labor force in rice farming and cultivated land which is used as a measure of land abundance. Even given the uncertainties and margin of errors inherent in such an exercise, the picture that emerges is striking and useful. Data show that, on the whole, public investments in irrigation have followed the movements of the relative scarcity of cultivated land per worker.

From the beginning of the twentieth century up to the 1950s, the amount of land devoted to rice increased four-fold. Over the same period population grew from about 8-9 million to just under 20 million. However, the annual rate of growth of land area remained substantially higher than the rate of population growth so that resulting densities remained relatively low and the land cultivated per worker increased continuously. From the 1960s until the mid-1980s, population pressure on land increased significantly, making land-saving technologies more attractive. Between 1960 and 1985, while the planted area per worker dropped by about 0.5-1.6% per year, irrigated areas grew between 2% and 5.4% per year; double-cropped areas also increased dramatically between 15-20% per year. Later on, starting in the mid-1980s, the growth rate of rice labor force began to slow down, due to a sharp decline in the natural rate of population growth and shifts from agricultural to industrial employment. This slowdown has reflected in a lower rate of decrease of planted area per worker, and an increasing contraction of land-augmenting investments.

The development of irrigation infrastructure also came as a direct or indirect response to external demand for rice output. In the post-World War II period, Thailand's international environment changed dramatically. Increased exports of Thai rice were needed to solve the postwar food crisis. The further acceleration in irrigation investments between 1975 and 1985 represented a response to the increase in the benefit-cost ratios for irrigation investment that was associated with dramatic increases in world rice prices. Since the mid-

1980s, the downward movement in the real world rice prices and the rising doubts about irrigation performance has led to some changes in the budgeting of investments in irrigation.

Regional differentials in population pressure does not appear closely linked to investments in irrigation infrastructure. Agricultural expansion in Thailand has been achieved by a gradual, proximate expansion of existing agricultural land so that, over time, the distribution of population has remained balanced. Investments in irrigation, by contrast, have been heavily biased toward the Central Region. Economic factors have been important in determining the location of irrigation investment. To harness the flooding of large Thai rivers over large extents of flat land required large capital investments. As soon as good locations for irrigation were developed, the construction costs for new large-scale irrigation systems increased. Other powerful forces included access to urban and foreign markets, which depends on the existence of transport infrastructure and marketing facilities well developed. The concentration of irrigation projects in Central Plain was also motivated by its closeness to Bangkok - the main export center - and its well endowment in transport and marketing facilities.

3.1.1.4 Overall Irrigation Performance and Water Management Problems

Irrigation performance is notoriously ambiguous and difficult to conceptualize and measure. Three dimensions of irrigation performance can be identified: the condition of the physical structure of irrigation systems, the water delivery effectiveness and the agricultural productivity of the systems. In all these three dimensions, most irrigated systems in Thailand have achieved a low performance, and it is widely held that existing infrastructure has not even insulated the irrigated segment of Thai rice farming against drought.

The irrigation technology choice has been essentially based on land characteristics (soil permeability and field slope), and water source (groundwater, surface water or both). Given the diverse physical characteristics in disparate locations in the country, there have been diverse means by which water delivery has been conducted. The large expanses of flat land in Central Plain made it possible for a single source to irrigate a relatively large area. In the North diversion dams were constructed. In more arid areas like the North East, the highly erratic rainfall pattern coupled with the very limited capacity of the soil to retain water made small-scale storage facilities much more relevant; a new type of irrigation, tank irrigation, was implemented in this region.

The irrigation scheme in Thailand was aimed primarily at water control in order to stabilize and increase yields of the wet season rice production. Therefore, most of the existing networks provide water only for supplementary irrigation of wet-season rice. In addition, the overall system is reportedly impeded by the uncertain and variable amount of available water because the supply of water depends very much on annual monsoon variability. As a result, the potential of the system has not been fully realized and the existing system does not have the capacity to support a revolution in production techniques (Ishii, 1978).

Water use and allocation has become increasingly conflictual as the resource progressively dried up. The level of water in the two major storage dams, the Bhumipol and the Sirikit, has decreased from 10 to 7 millions m3 in the last twenty years. Irrigated area expanded from nearly 1 million rai in the 1970s to 6.5 millions in 1999, and industrial and urban consumption doubled during the last ten years. In practice, urban households have been given priority in the state policy regarding water allocation, so that farmers have been largely left out in recent years, particularly when the rains have been poor[26]. Consequently, the water supplies from the irrigation network are inadequate in relation to demand.

Finally, the extent to which the potential of irrigation structures can be realized depends upon whether effective institutions exist that provide social capital in terms of managerial capability and productive working relationships to support the maintenance of the infrastructure. The construction of physical capital and the development of social capital are intricately related. How individuals work with one another significantly affects irrigation performance.

Water management in Thailand is highly bureaucratic and centralized. There are currently thirty-eight government agencies under seven ministries dealing with water. However, water supply and regulation is under the responsibility of three main institutions: the Royal Irrigation Department (RID) has the responsibility of operating and maintaining irrigation systems as well as supplying irrigation water to the agricultural sector; the Electricity

[26] The drought in 1997, for example, resulted in more than four million rai of agricultural land damaged in irrigated areas.

Generation Authority of Thailand (EGAT) is responsible for water delivery in the storage dams; the Metropolitan Waterworks Authority (MWA) supplies water to Bangkok. Because of little coordination between these agencies and the resulting overlapping interventions, there are many who argue that water problems stem not from a shortage of water but rather from its inefficient use (Christensen and Boo-Long, 1994; Kaosa-ard and Kositrat, 1993).

Peasant participation in this system is minimal and the persistent under-performance of irrigation agencies has raised serious doubts about the usefulness and feasibility of relying on government officials to manage irrigation services. Effective water delivery largely hinges on farmers following water allocation rules, taking care of the system daily, and contributing their efforts to the maintenance of the structures when necessary. How much and how well farmers mobilize adequate resources for operation and maintenance affects the overall performance of irrigation system. Without this participation, the technicians are unable to implement the various systems for controlling water volume, timing the water supply and organizing rotational irrigation. As a result, an « open-access » water regime prevails which means that any user can freely draw as much water as needed from rivers and irrigation canals.

3.1.1.5 Cropping Intensity

In the early 1960s, a ten-year plan for promotion of dry season agriculture was drawn up. Up to the mid-1970s, the expansion of the dry-season paddy exceeded the planning targets. Double cropped area grew by 10% every year in the first half of the 1960s, by 15% in the second half and by 20% in the first half of the 1970s (cf. Table 3.2). This growth rate slowed considerably from 1975 to the end of the 1980s. During the first half of the 1990s, double cropped area decreased first significantly and had subsequently a very moderate rate of increase. Expressed as a percentage of total area cultivated, double cropped area covers only some 10% of the wet season acreage (less than 10% in the central region, 2.5% in the North, and less than 1% in the Northeast).

Among the factors responsible for the recent weak progression of double cropping at the national level, we find the continued growth of urban water demand coupled with the virtual halt of water supply development. Because increasing proportions of available water are being used for industry and to supply the needs of growing urban areas, the government has tended to reduce water supply to farmers for the dry season crop, and in some areas dry season water is rotated from year to year. Irrigation is given relatively low priority in water uses, and farmers are forced to bear the burden of reduced diversions necessary to meet increasing urban demand. Therefore, despite repeated assurance from the RID that there will be adequate water for the six to seven million rai of off-season rice fields, the decrease in dry season rice cropping that occurred until 1990 is likely to continue in the future due to the lack of a proper water supply.

Table 3.2: Trends in Double Cropped Area

	Annual growth rate Double cropped area	Percentage of double cropped area
1960-1965	10.3	0.2
1965-1970	15.4	0.9
1970-1975	20.4	3.3
1975-1980	5.7	5.3
1980-1985	4.3	6.4
1985-1990	8.7	7.2
1990-1995	-4.6	6.5
1995-1999	3.7	9.7

SOURCE: Office of Agricultural Economics

The availability and cost of labor may also have been a significant barrier to double-cropping expansion. As discussed in Chapter 2, the yield gains from the adoption of double cropping technology come at the expense of increased family and hired labor use. When farmers cultivate two crops in a year, labor devoted to rice farming increases sharply. At certain levels of industrialization and thus of wage levels, the cost of maintaining a double cropping system in terms of time (and income) lost from factory work, more than compensate the production increase. As the opportunity cost of labor has sharply risen in Thailand during the 1980s in response to the overall economic growth, producers' resistance to double cropping has increased and total household income has been more and more supplemented by off-farm employment earnings.

It is further interesting to relate, at the regional and microeconomic level, the number of crops grown within a year and factors such as the presence and the quality of an irrigation network, and other factors among which farm size appears relevant. On the whole, average cropping intensities at the provincial level are consistent with the percentage of area irrigated. Considering our sample, irrigated area constitutes about 93% of the area cultivated in Suphan Buri. The area covered by the public irrigation network is more limited since only 42% of the cultivated area has a continuous water supply through dam irrigation. Pumping irrigation covers another 41% while the area covered by both gravity and pumping irrigation is 18%. The average cropping intensity index is 171. In contrast, all farmers in Pitchit have not benefited from the establishment of irrigation systems; 85% of cultivated area appears to be under some kind of irrigation benefits. Accordingly, the average cropping intensity index is 137. Finally, irrigated rice farming is inexistent in Roi Et so that only one crop a year is possible.

Interestingly, small farmers have a larger proportion of their cultivated area which is irrigated compared to large farmers. The main reason why, in a given province, smaller farms hold a higher percentage of irrigated land derives from the nature of land that passes hands in the land market arising from the prevalence of distress sales (cf. Chapter 1). Since farmers first depend on land resources as their main source of employment, income and food, those who are compelled by circumstances to sell a part of their land usually part with the inferior pieces of agricultural land, or with pieces of land best suited for nonfarm uses, which are in both and most cases non-irrigated land. Therefore, the buyers of land, that is large farmers, are always somewhat forced to buy and cultivate lower quality pieces of land.

Table 3.3: Source of Irrigation and Cropping Intensity by Province and by Farm Size

	< 15 rai	16-30 rai	31-50 rai	51-100 rai	101+ rai	All sizes
Suphan Buri						
% of area irrigated	100.0	96.3	98.2	83.6	50.0	93.0
by dam irrigation	42.0	46.1	32.3	27.9		41.6
by pumping irrigation	43.6	39.8	45.2	41.3		40.6
by both	14.4	14.1	22.5	30.7		17.7
Cropping intensity	172	176	164	167	100	171
Pitchit irrigated						
% of area irrigated	100.0	85.7	95.4	79.2	76.1	85.3
by dam irrigation	16.0	31.6	29.3	36.6	79.0	40.1
by pumping irrigation	84.0	68.4	59.9	46.9	21.0	51.8
by both	0.0	0.0	10.8	16.5	0.0	8.1
Cropping intensity	150	152	135	113	105	137
Pitchit rainfield						
% of area irrigated	0.0	0.0	0.0	0.0	0.0	0.0
Cropping intensity	100.0	100.0	100.0	100.0	100.0	100.0
Roi Et						
% of area irrigated	0.0	0.0	0.0	0.0	0.0	0.0
Cropping intensity	100.0	100.0	100.0	100.0	100.0	100.0

SOURCE: ORSTOM/CUSRI Project, Survey on Rice Farmers.

3.1.2 Adoption of Improved Varieties

The adoption of high yielding varieties (HYVs) that began in the 1960s in the developing world has progressed steadily over the past three decades. These varieties were introduced in Thailand only in the late 1970s, several years later than in the neighboring countries. At that time, it was widely believed that rice production in Thailand was just about to undergo a green revolution as a result of the introduction of higher yielding varieties. Today this optimism has all but vanished because the introduction of these varieties has not become as widespread as was originally hoped. Although HYVs can undoubtedly make important contributions to increased productivity through their land-saving component and through the more efficient use of inputs, the rate at which Thai farmers have responded to their availability has been very low.

Several factors explain this lack of enthusiasm. First, HYVs are unsuitable in rainfed agricultural areas which constitute most of the rice growing ecological environment in Thailand. HYV seeds have low tolerance limits for water stress and adverse output effect can be considerable in case of drought. In contrast, the area adoption has rapidly increased where these varieties were suited to the local natural environment; in Central Plain, nearly 40% of the wet season rice area is sown with HYVs. Therefore, the further spread of HYVs in Thailand certainly depends on the success in developing varieties suitable for marginal areas or on better water control development.

Second, other factors than water availability have been at work. The absence of intensification in Thai rice farming can be considered as a response to the type of development that has occurred in manufacturing; the characteristics of agricultural technology adopted were conditioned by this. Since the manufacturing sector has been highly labor-demanding, labor bottlenecks and increasing labor costs have been important constraints to the adoption of labor-intensive cultivation methods. Although the yield advantage of HYVs over traditional varieties can be considerable in irrigated areas, this technology has been perceived by farmers as unattractive given the increasing cost of labor.

Third, although it is usually assumed that adoption of HYVs is constrained by supply factors, demand factors have certainly played a crucial role in Thailand. The late introduction of new varieties in the Kingdom was partly due to the fear by the authorities that the grain quality of these varieties was inferior to the traditional Thai varieties and that export markets might be lost. This fear has proved realistic. Thailand is currently losing its advantage in low quality rice exports due to rising production costs and emergence of new competitors. Vietnam is in the process of improving the productivity of its rice sector and has become an important rival to Thailand on the international market. The recent spurt in Vietnamese rice exports as a result of economic liberalization has forced down the price of low rice qualities in the world market. Stiff competition from other rice exporters such as India, Pakistan and China have further strengthen this trend. Even if Thailand can take the opportunity of the recent depreciation of the baht to accelerate rice exports and capture as many potential markets as possible, Thailand's chances of preserving its ranking as the world's largest rice exporter depend on its ability to specialize in higher quality products in order to cope with this competition.

The high quality, aromatic Thai rices are definitely low-yielding, traditional varieties. Since the output price of rice has always been allowed to fluctuate, quality differences have traditionally been reflected in the price on both international and domestic markets[27]. Fragrant rices command high prices on both markets. Therefore, in marginal areas, where the yield advantage of HYVs is low, quality traits of traditional varieties have largely outweighed the yield advantages of HYVs and have limited their adoption. There is therefore no evidence to justify efforts to develop HYVs in these areas.

The Thai Department of Agriculture has conducted research to develop varieties more targeted to specific environments. Desirable traits other than high yields have been developed through local breeding programs. In irrigated areas the major objective has been to develop earlier maturing varieties, which, in turn, have allowed increased cropping intensity. Another line of research has been to improve pest resistance, permitting reductions in the use of pesticides and promoting greater stability of yields. Recent investigations

[27] For example, in 1997, the export price of high quality rice such as fragrant rice was US$ 740 while the price of 100% grade and 5% grade was US$480 and $380 per ton, respectively. 15% grade rice was exported for an average price of $340, 25% grade was $320, while the export price of super broken rice was $237 per ton. Accordingly, the price of rice received by producers ranged from 3,900 baht a ton for low quality rice while fragrant rice hit a record of 10,000 baht per ton.

by the Department of Agricultural Extension shows that 76% of farmers surveyed use pesticide sprays for the wrong pests and apply them at the wrong time. These applications are therefore unnecessary. Moreover, farmers complain about the negative effects on the health of people exposed to pesticides. Improved pest resistance of new varieties have contributed not only to increased yield stability but also to enhancing farmers' health.

The rate of adoption of resistant varieties has been quite successful. The benefits of seeds that are resistant to drought, disease, and pests are generally independent of land and labor costs. The main advantages of these seeds is that they do not require any extra labor while allowing a lower application rate of pesticides. Our survey indicates a higher rate of adoption in the rainfed areas in the North (66.7% of farms) and Northeast (71.8% of farms) where farmers do not spend much on pesticides (cf. Table 3.5). Conversely, in the Central Region, where the adoption of pest-resistant varieties has been lower, the use of pesticides is most widespread. In this area, continuous cropping throughout the year indirectly aggravated the buildup of pest populations and exacerbated pest losses.

3.1.3 Fertilizer Use and Factor Prices

3.1.3.1 Trends and Patterns in Fertilizer Use

In 1965 manure and compost were widely used to maintain fertility, as they had been for generations, and synthetic fertilizer use was not present. Between 1965 and 1995 the fertilizer application rate on the major crop rose from 30.0 kg to 106.6 kg per cultivated hectare, while the dry season application rate rose from 142.8 kg to 320.7 kg per hectare. These increases represented an annual average growth in fertilizer use comprised between 8% and 12%. However, it is estimated that fertilizer use in Thailand is approximately 16% of the theoretical optimum level of fertilizer use (FADINAP, 1995). Therefore, despite rapid expansion of fertilizer use, Thailand is among the lowest consumers of chemical fertilizer in Asia and the average fertilizer application rates are still far behind those countries which are considered as optimal users.

Table 3.4: Trends in Fertilizer Consumption

	Annual growth of fertilizer use	Fertilizer use (kg per ha) first crop	second crop	Output/input price ratio
1965-1970	0.5	30.0	142.8	0.57
1970-1975	4.4	24.8	163.7	0.51
1975-1980	10.8	31.8	233.7	0.70
1980-1985	8.2	43.1	296.0	0.68
1985-1990	12.3	63.3	310.0	0.79
1990-1995	11.1	106.6	320.7	

SOURCE: Office of Agricultural Economics

A common observation from farm household surveys in developing countries is that many households do not apply any chemical fertilizer. At the time of our survey, chemical fertilizers were used by almost all farmers (85-100%). However, levels of fertilizer use differed widely between provinces and within provinces between households. The Central region consumes more fertilizer than any other region. The application rate in Suphan Buri (391 kg per ha) is far higher than the sample average of 214 kg per hectare of rice. In the irrigated areas of Pitchit, average fertilizer consumption is quite low. Although the percentage of farmers using organic fertilizer in the North East is high, the average fertilizer consumption in Roi Et is only 160 kg per hectare. The recommendations for fertilizer use vary between 215 kg/ha and 294 kg/ha, depending on the paddy variety, the type of soil and the material used. While the province of Suphan Buri is within the recommendations, in Pitchit and Roi Et the quantities applied are only half of the recommended fertilizer rates.

It is difficult to determine the reasons for the low national fertilizer consumption and the wide variations observed among provinces and farmers, but some speculation can be offered. The optimum level of fertilizer use for the farmer is first determined by the fertilizer response function. Crop response to current input application is, in turn, determined by a large number of factors such as crop variety, irrigation, soil-type, kind of material used, knowledge, experience and management. The second set of factors that may influence fertilizer levels are associated with the choice of input level in an uncertain environment when farmers are risk-adverse. The fertilizer intensity application depends crucially on the relationship between risk aversion and wealth, which is usually significantly correlated with farm size and tenure. Third, in deciding the amount of input use the farmer is generally guided by the prices of the agricultural products he will harvest compared to the costs of inputs.

3.1.3.2 Agro-Environment Constraints on Fertilizer Use

The first argument usually put forth to explain levels of fertilizer use is the percentage of cropped area that is irrigated. Since a greater assurance of water supply reduces yield fluctuations and increases the potential of yield rates as well as the productivity of inputs, irrigated areas are likely to adopt fertilizer faster than rainfed ones. Some authors have argued that the process of innovation is sequential and the adoption pattern is hierarchical (Rauniyar and Goode, 1996), that is the probability of adopting a practice is conditional on adopting practices that precede it in the sequence. In this case, irrigation would be a prerequisite to other intensive practices.

Table 3.5 shows substantial differences in the level of fertilizer application arising from differences in the stage of irrigation development. Farmers adapt their cropping behavior according to the risks they face and uncertain rainfall combined with poor irrigation act as major deterrents to fertilizer use. In the Northeastern region, overwhelmingly rainfed farming confronted with a highly variable rainfall pattern, intensification is risky. Because fertilizer can burn crops on dry soils, northeastern farmers may hesitate to apply fertilizer to their fields. Another reason is that under the prevailing soil and climatic structure increased use of fertilizers would only allow yield increases at a decreasing rate of profitability. In contrast, a significant irrigation capacity has been provided for farmers in Central Plain. The high degree of fertilizer use is therefore not surprising. The importance of water control is further confirmed by the fact that the per-hectare application of chemical fertilizer is much higher on the second than on the major crop because the dry season crop is grown under more predictable environmental conditions on irrigated land.

Table 3.5: Seeds, Planting Technique and Current Input Use by Holding Size

	<15rai	16-30mi	31-50rai	51-100rai	101+rai	All sizes
Suphah Buri						
Use of recommended seeds (% of farms)	53.1	47.6	68.2	33.3		53.0
Planting technique (% of farms)						
Broadcasting	100.0	100.0	100.0	100.0	100.0	100.0
Expenses on herbicides						
% of farms	96.9	97.6	90.9	100.0	100.0	96.0
Expenses on first crop (baht/ha)	433	306	331	346	188	352
Expenses on second crop (baht/ha)	377	353	403	406		372
Expenses on pesticide						
% of farms	87.5	97.6	95.5	100.0	100.0	94.0
Expenses on first crop (bath/ha)	608	387	422	342	188	462
Expenses on second crop (bath/ha)	568	349	376	244		422
Expenses on Fertilizers						
% of farms using fertilizer	93.7	97.6	95.5	100.0	100.0	96.0
Application rate first crop (kg/ha)	460	407	382	421		391
Application rate second crop (kg/ha)	455	41'2	385	381		419
Pitchit (irrigated areas)				71.4	75.0	63.5
Use of recommended seeds (% of farms)	78.6	71.4	38.1			
Planting technique (% of farms)						
Broadcasting	42.9	52.4	47.6	21.4	50.0	43.2
Transplanting	42.9	23.8	38.1	50.0	25.0	36.5
Broadcasting and transplanting	14.3	23.8	14.3	28.6	25.0	20.3
Expenses on herbicides						
% of farms	50.0	81.0	85.7	85.7	25.0	74.3
Expenses on first crop (bath/ha)	113	102	77	52	20	83
Expenses on second crop (bath/ha)	37	92	185	117	375	118
Expenses on pesticide						
% of farms using pesticides	92.9	76.2	71.4	78.6	75.0	78.4
Expenses on first crop (bath/ha)	333	102	130	98	109	153
Expenses on second crop (bath/ha)	162	365	299	196	0	266

Expenses on Fertilizers						
% of farms using fertilizer	78.6	95.2	81.0	92.9	75.0	86.5
Application rate first crop (kg/ha)	176	139	113	101	64	127
Application rate second crop (kg/ha)	208	221	230	308	250	234
Pitchit (rainfed areas)						
Use of recommended seeds (% of farms)	80.0	25.0	87.5	66.7		66.7
Planting technique (% of farms)						
Broadcasting	20.0	50.0	50.0	16.7		33.3
Transplanting	80.0	25.0	37.5	83.3	100.0	58.3
Broadcasting and transplanting		25.0	12.5			
Expenses on pesticide (bath/ha)	15	28	44	166		138
% of farms using pesticides	20.0	50.0	37.5	50.0		41.7
Expenses on herbicide (bath/ha)	54	36	52	165		150
% of farms using herbicides	80.0	50.0	87.5	66.7		75.0
Application rate of fertilizer (kg/ha)	80	78	90	77		83
% of farms using fertilizer	60.0	25.0	75.0	66.7		58.3
Roi Et	66.7	77.3	80.0		100.0	71.8
Use of recommended seeds (% of farms)						
Planting technique (% of farms)	1.7		5.0		100.0	2.9
Broadcasting	98.3	100.0	90.0			96.1
Transplanting			5.0			1.0
Broadcasting and transplanting	20	15	30		0	21
Expenses on pesticide (bath/ha)	45.0	77.3	80.0			58.3
% of farms	1		2		32	1
Expenses on herbicide (bath/ha)	1.7		5.0		100.0	2.9
% of farms	186	143	120		59	163
Application rate of fertilizer (kg/ha)	98.3	95.5	100.0		100.0	98.1
% of farms using fertilizer						

SOURCE: CUSRI/ORSTOM Project, survey on rice farmers

The lack of fertilizer-response of most Thai varieties is another argument which explains average low fertilizer application rates in Thailand. The decision to use fertilizer is concurrent with the decision on whether or not to plant HYVs. A crucial characteristic of HYVs is their high yields when used in conjunction with cemical fertilizer. The likelihood of adopting high yielding varieties and fertilizer together are mutually conditioning because of the interactions between these practices. These two practices generally occur simultaneously and they are usually adopted as a package or not at all. In Northeastern and Northern part of Thailand, glutinous rice, the low yielding traditional subsistence crop of farmers, is cultivated on, respectively, 72% and 37% of rice fields. There is ample evidence that this local Thai variety is not very responsive to increased fertilizer application. The high quality jasmine rice, grown in the lower areas of Northeast, and which may represent the main comparative advantage of Thai rice in the international market, is also a low yielding traditional variety. Moreover, soil scientists argue that jasmine rice requires unique geographical characteristics such as dry weather and salty soil and that it will only be able to maintain its unique flavor if it grows naturally without any special input such as fertilizer.

Finally, it is possible that fields in some regions such as Roi Et benefited from manuring from other sources (livestock) and are thus less likely to receive chemical fertilization. Since households have manure resources that are largely determined by the size of their livestock, animal ownership is expected to exert a negative influence on fertilizer use. However, in our data, fertilizer application rates between owners and non-owners of buffaloes differ only slightly.

3.1.3.3 Size and Tenure-Constraints to Fertilizer Use

Although fertilizer use is regarded as being neutral to scale, constraints such as those arising from access to extension services, lack of knowledge about the productivity of fertilizer, marketing problems are expected to bias fertilizer use towards larger farmers. Furthermore, since larger farms enjoy preferential access to credit, it follows that cash inputs which are credit-intensive should be less readily available to small farmers. Hence, fertilizer intensity should be positively correlated with farm size.

The evidence of higher intensity of fertilizer use on large farms is neither clear-cut nor easy to interpret. Some relevant findings summarized in Table

3.5 suggest that actually small farmers have had relatively good access to fertilizer and that they consequently have been adopting biological technologies as well as large farmers. Although the percentage of farms using fertilizer tends to increase with holding size, fertilizer is used more intensively by producers farming less land, while farmers with bigger tracts of land tend to apply smaller doses to larger areas. For example, the fertilizer use was 460 kg per hectare for small farms and 421 kg/ha for large ones in Suphan Buri, 138 kg/ha and 48 kg/ha, respectively in irrigated areas of Pitchit, 183 kg/ha and 60 kg/ha in Roi Et.

We can interpret this result as evidence of differences in relative risk exposure. We have observed that the probability of having inadequate irrigation is positively related to farm size; small farms have a higher percentage of area under irrigation as evidenced in Table 3.2. The comparative advantage of small farms, reflected in endowments of factors such as irrigation facilities which are used intensively in the production of rice, and which reduce the variability of returns, explains the higher fertilizer use by smaller farms. Larger farms compensate for the higher risk by reducing the levels of variable input such as fertilizer. This reinforces the importance of irrigation in increasing the consumption of fertilizer.

A question which has been frequently addressed is the implications of land tenure for technological innovation. It is the conventional view that tenancy retards the progress of technological innovation. Table 3.6 presents the proportion of farmers who have adopted fertilizers and the amount of application by tenurial status. Although chemical fertilizers are applied by almost all farmers irrespective of land tenure status, the application rate of fertilizer by tenants and part owners is generally lower from that by owner-operators. This suggests that tenurial status is a relatively significant determinant of the rate of fertilizer application[28], and that land tenure insecurity tend to discourage tenants from working toward a higher productivity.

[28] However, the best way of verifying factors like tenurial status which affect farming technique would be to compare input intensities on owned and leased plots farmed by the same individual. Unfortunately, we could not separate from available data the quantity of fertilizer each field received.

Table 3.6: Fertilizer Use by Province and by Tenure

	Landlord owners	Owners	Part owners part renters	Tenants	All tenure
Suphan Buri					
% of farms using fertilizer	100.0	97.3	92.0	96.0	96.0
Application rate of first crop (kg/ha)	441	473	410	353	425
Application rate of second crop (kg/ha)	465	461	421	348	419
Pitchit irrigated					
% of farms using fertilizer	75.0	91.3	75.0	87.5	86.5
Application rate of first crop (kg/ha)	132	137	88	137	127
Application rate of second crop (kg/ha)	208	249	181	155	234
Pitchit rainfed					
% of farms using fertilizer	100.0	54.5	33.3	83.3	58.3
Application rate of first crop (kg/ha)	93	60	105	100	83
Roi Et					
% of farms using fertilizer	100.0	98.7	93.3	100.0	98.1
Application rate of first crop (kg/ha)	234	164	129	118	163

ORSTOM Project, Survey on Rice Farmers

3.1.3.4 Input/Output Price Ratio

Finally, the demand for fertilizer is expected to depend to a large extent on the relation between crop and fertilizer price; a low profitability of chemical inputs have repercussions on their use. Thailand has been importing around 85% of its chemical fertilizer requirements. Approximately 90% of fertilizer imports have been handled by the private sector. The largest public sector importer, the Marketing Organization of Farmers, started operations in 1974 with the intention of supplying farmers with low-cost inputs. In 1981 BAAC started procuring its own fertilizer and later developed a credit-in-kind system of supplying fertilizer.

Despite price reductions of nearly 20% compared to the market price during the second half of the 1970s and 7-8% since the 1980s, the public sector failed to significantly affect farmers' consumption owing to its limited scale of operation[29]. The MOF supplied approximately 50% of the total fertilizer requirements to paddy farmers during the first half of the 1980s, but this share decreased significantly over time. Over the last decade, the public sector as a whole (MOF and BAAC) channeled no more than 27% of the fertilizer consumed in Thailand, of which MOF sales represented slightly more than 10% of total fertilizer consumption (Office of Agricultural Economics). Therefore most chemical fertilizers required by farmers have been purchased from the private retailers.

Price incentives for using chemical fertilizer have been weak for another reason. Price movements on the world market have been passed along directly to the local market. As a result, fertilizer prices paid by Thai farmers were closely related to the import prices[30]. During the sixties, because of stability in both input and output markets, very little variation in the input/output ratio was observed. C.I.F fertilizer prices did not change much and this period was also characterized by remarkable stability in world rice markets and by low rice prices. Added to that, the Thai rice bore an overall taxation of nearly 40% of the border price (World Bank, 1980), which produced a strong disincentive

[29] An obstacle to increased use of MOF fertilizer was the fact that most MOF warehouses were not located throughout the country. Because of many delays in fertilizer delivery, supply has not always coincided with farmers' needs during the peak season. Adulteration of fertilizer has been another serious problem for many years and has had a marked negative effect on fertilizer consumption (FADINAP, 1990).

[30] The large number of importers attests to the fact that access to the Thai fertilizer import market is not hampered by major institutional or tariffs barriers. In addition, importers' price margins have been relatively low due to the intense competition among traders (TDRI, 1992).

to use fertilizer. Upward movements in both rice and fertilizer prices were then pronounced under the impact of the first and second oil crisis. Because rice prices increased more than fertilizer prices the input/output ratio improved significantly over short periods of time, but then deteriorated rapidly. During the 1980s, in the face of depressed world rice prices, there has been a considerable reduction in rice export taxes. The rice premium was removed for good in 1986 so that at the present time, there is minor government interference in the rice market. As a result, the price ratio of paddy to compound fertilizer improved significantly. Unfortunately, the recent sharp fall in the value of the baht has caused the price of chemical fertilizer to increase again by nearly 30%. In addition, most regional dealers have closed their business because the lines of credit usually given by importers have been revoked due to tight liquidity. All of this will certainly discourage farmers to use growing quantities of fertilizer.

The differential in the fertilizer/output price ratio can be expected to explain the disparity in fertilizer use at the local level. The ratios of fertilizer to farmgate paddy prices with respect to regions and farm size are shown in Table 3.7. Both fertilizer and paddy prices paid/received by farmers vary widely, depending on the source and the terms on which the fertilizer/paddy is bought/sold. The prevailing market prices for rice and fertilizer implies that an average of 1.5 kg of crop has to be sold to earn 1 kg of fertilizer. Price relationships of this order are not conducive to high fertilizer use. However, the differences between regions are significant: the Northeastern province faces the lowest incitement to fertilizer consumption, whereas the input/output ratios in Suphan Buri and Pitchit are more encouraging. The regional disparity in the structure of incitements to fertilizer consumption is thus consistent with disparity in regional fertilizer use, lower applications of chemical fertilizers corresponding with less stimulating grain/fertilizer price ratio.

Table 3.7: Fertilizer/Paddy Price Ratios

	< 15 rai	16-30 rai	31-50 rai	51-100 rai	101+ rai	All sizes
Suphan Buri						
Fertilizer price (baht/ton)	3,312	3,378	3,514	3,400		3,382
Paddy price (baht/ton)	5,369	5,053	5,195	5,087		5,182
Input/output ratio	1.62	1.50	1.48	1.50		1.53
Pitchit						
Fertilizer price (baht/ton)	3,350	3,400	3,622	3,929	3,920	3,604
Paddy price (baht/ton)	5,043	5,126	4,965	5,241	5,600	5,105
Input/output ratio	1.51	1.51	1.37	1.33	1.43	1.43
Roi Et						
Fertilizer price (baht/ton)	3,358	3,312	3,558		3,900	3,411
Paddy price (baht/ton)	5,730	5,743	5,580		5,000	5,696
Input/output ratio	1.71	1.73	1.57		1.28	1.67

SOURCE: ORSTOM/CUSRI Project, Survey on Rice Farmers.

Larger farmers tend to obtain better prices for the rice they sell while they tend to purchase fertilizer at lower prices than small farmers. According to the marketing part of our survey, large farms are in a position to get higher prices for their production than small ones for two main reasons. First, large farm sales usually lag one or two months behind small farms. This reflects the greater latitude bigger farmers have about the choice of the time of sale owing to their greater « *holding capacities* » defined by storage space, cash requirements, and wealth. Second, the same trader tends to offer different prices to different classes of producers and smaller producers always receive lower prices than larger producers whatever the marketing channel used. If it is clear that small farmers are too experienced to be cheated on a large scale by traders or rice mills, their bargaining power seems to remain weak.

Furthermore, if we take into account the fact fertilizer is usually bought on credit, the price position of small farmers worsen. The terms and conditions of credit contracts vary with the source of finance (cf. Chapter 4). The BAAC has traditionally been the major institutional supplier of credit for fertilizer and charge an annual interest rate between 9% and 12%. In contrast, local merchants charge an interest rate of about 3% to 5% per month. Since small farmers encounter difficulties in securing institutional credit, buying fertilizer on credit is much more expensive than buying fertilizer for cash. This tend to reinforce the price position of large farmers.

However, despite large farmer paddy sales generally fetch better prices, they do not tend to invest more liberally in fertilizer than smaller ones. One rationale behind this behavior is that limited water control on the parcels cultivated by large farmers makes use of fertilizer economically risky. By contrast, the tendency for irrigated land to be concentrated in the smaller holdings tends to outweigh the adverse price relationships faced by small farms. A second rationale has largely been discussed in Chapter 2. An increase in fertilizer use lead to an increase in farm labor use and thus labor cost that is far more greater than the benefits reaped from larger fertilizer application. It is therefore doubtful whether large households will respond to higher output prices or lower fertilizer prices by applying more fertilizer.

3.1.4 Planting Technique and Herbicide Use

There are two principal methods of rice seeding: direct seeding (dry or wet) or broadcasting, and transplanting. A large variation can be observed in the method of rice seeding between provinces. The survey data record a decline in the ratio of sampled farmers broadcasting rice from 100% in Suphan Buri, to 40.8% in Pitchit, and less than 3% in Roi Et. The choice of one rice establishment practice over another is governed by several factors such as hydrological conditions, production technology, and relative factor prices.

First, planting techniques differ substantially for rainfed and water controlled agriculture. The proportion of the area broadcast roughly corresponds to the proportion of land under irrigation. Broadcasting is best suited for areas where irrigation and drainage conditions are adequate because this technique requires careful leveling of fields and good control of water depth. Seeds are broadcast directly into the main paddy fields and the water level is increased in line with the growth in the seedlings. In contrast, transplanting is a widely-practiced traditional method of rice seeding in rainfed areas, because it reduces the risks associated with rainfall uncertainty at the beginning of the season. Farmers transplant only after enough rain has fallen to facilitate land preparation and plant growth.

Second, from a technical point of view, the availability of shorter duration cultivars coupled with direct seeding improve the feasibility of higher cropping intensities where agroclimatic conditions are propitious. Double cropping reduces the turnaround time between the first and the second crop, which makes impossible transplanting the first crop. Broadcast rice is harvested earlier than transplanted rice so that the subsequent crop can be grown in the given time.

Third, a major technical shortcoming of the direct seeding method is that it provides a more favorable environment for weed growth since crop and weeds emerge together. Direct seeding thus calls for a substantial increase in the use of herbicides. Accordingly, chemical weed control must be a viable and affordable technology. In contrast, transplanting makes weeding easier and more efficient where water supply cannot be controlled because weed growth is hampered by the presence of standing water in transplanted plots. Herbicide use has therefore closely been related to the introduction of broadcasting in the Central Region as shown in Table 3.5[31]. In the Northeast herbicides are used to a limited extent in rice farming.

Fourth, rising labor costs also have strong repercussions on the planting technique. Broadcasting and transplanting process requires widely divergent quantities of labor. Broadcasting requires on average half of the hired labor used for transplanting, and 10% to 25% less of the family labor (cf. Chapter 2). Since direct seeding must be combined with adequate chemical weed control to be effective, the relative profitability of the two techniques is primarily a function of the relative price of labor and herbicides. Agricultural wage rates in Suphan Buri were found to be roughly 1.8 times the level of Pitchit, which in turn exceeded that of Roi Et by nearly 15%. The price of herbicides appeared roughly equal between provinces. It is therefore not surprising that the adoption of direct-seeding has rapidly spread throughout the irrigated areas of Central Plain in response to rising labor costs. In areas where agricultural wages are still low, the cost of buying herbicides is still greater than any saving in labor costs that may result from a shift from transplanting to direct seeding, and the farmers have little incentive to adopt this labor-saving technology.

3.2 The Success of Mechanization

Until the 1950s, agricultural capital in Thailand was produced largely within the agricultural sector through labor-intensive methods. Traditional capital investment took the form of the basic tools usually made of wood (plows with iron shares, harrows, spades, hoes, sickles, flails), seeds, several types of irrigation pumps operated with wind and draft animals (Feeny, 1982; Ingram, 1971). Human and animal labor was virtually the only power source. The introduction of factory-made capital into rice farming has only taken place on a substantial scale since the early 1960s and nowadays many forms of capital used are produced in the industrial sector. Not only irrigation but numerous agricultural tasks have been mechanized in a flow of innovation.

---

[31] Table 3.5 refers to the expenses on herbicides per unit of area. The wide heterogeneity in the unity of measurement made the estimation of application rate of this input impossible. Since the price structure does not fluctuate much between provinces, it is argued that the differences in expenses on herbicides properly reflect the differences in the volume used.

3.2.1 Trends in Mechanization

There have been unsuccessful attempts at farm mechanization in Thailand since 1910, mainly because the machines imported were expensive and not suited to local conditions. Since the 1960s, the extent and rapidity of the mechanization process has been striking. Table 3.8 presents estimates of the stock of farm machines in Thailand and major trends in mechanization over the 40 years between 1960 and 1999. National data on the number, composition, and use patterns of agricultural machinery have been available since 1950 for tractors and water pumps. Unfortunately, it has not been possible to separate those used in rice production[32]. We therefore have to assume that mechanization in rice farming has expanded at the same pace than mechanization of agriculture. In addition, these figures do not provide an indication of the degree to which the stock of mechanical power is utilized, that is how many horsepower per hectare machines do contribute to the total power needs of agricultural production. For all these reasons the figures presented in this section should be viewed with caution and interpreted as orders of magnitude rather than as precise estimates.

Table 3.8: Annual Growth Rates of Farm Equipment

	Tractors two wheels	Water pumps	Sprayers	Threshing machines	Buffaloes
1960-1965		23.5			2.0
1965-1970		23.5			2.3
1970-1975		7.5			1.4
1975-1980	14.7	10.4			1.1
1980-1985	8.5	6.5		8.1	0.7
1985-1990	12.6	11.6	14.9	6.5	-4.3
1990-1995	12.1	11.1	14.8	6.6	-2.3
1995-1999	16.1	11.2	13.4	13.1	-4.4

SOURCE: Office of Agricultural Economics

[32]There is an exception for tractors. Imports of tractors consisted mainly of four-wheel tractors which are almost exclusively used in upland agricultural areas while two wheel tractors are used primarily for paddy fields.

The sequence in which machines were introduced in Thailand appears to follow a generally predictable pattern with energy intensive operations such as tillage and water pumping being mechanized initially and most intensively (Binswanger, 1986). Mechanization started with the importation of water pumps in the mid-1950s. Meanwhile a water pump designed by the Division of Agricultural Engineering was produced by local manufacture and commercialized. Interest in water pumps among Thai farmers grew very fast as farmers tried to secure water supplies. Mechanized pumps were thus increasingly being used to supplement or replace the traditional irrigation. Over the period considered the number of water pumps has increased remarkably as indicated by the average annual growth rates.

The mechanization of land preparation came later. The rapid growth of the power tillers and two-wheel tractor use was a phenomenon of the late 1960s and the early 1970s. In the early 1950s, animals constituted the main source of farm power. Small farm tractors were introduced from Japan in the mid-1950s. In 1957, the Department of Customs recorded the import of 267 tractors. But these machines failed to be widely adopted primarily because of their high cost. At the end of the 1960s the local farm machinery industry began to simplify a two-wheel tractor imported from Japan and to adapt it to local rice farm conditions. The first Thai-made versions of two wheeled tractors appeared. These tractors were first marketed in 1965 and gained popularity due to their low price compared to imported machines and their ability to perform well under local conditions. Since this period, motors have ceased to be unusual items of equipment, and an industry of small-scale agricultural machinery producers and suppliers grew up to support their use. In the face of the considerable potential market, the number of firms manufacturing tractors grew from a few in the mid-1960s to more than 150 in the 1990s. Most of the firms are located in the Central region and in the greater metropolitan area of Bangkok, which provide both better infrastructure and more potential customers.

One reason for the success of Thai power tillers among rice farmers stems from the fact that their initial design and construction was the product of small agricultural machinery producers who worked closely with farmers. Many changes were introduced to suit customers' recommendations and needs. Therefore, the design and engineering of the power tiller improved steadily over time although the basic design remained unchanged. Moreover, local tractors were reasonably cheap and simple to operate and maintain owing to a system of a detachable, all-purpose motor.

However, despite the number of power tillers in existence has increased quite dramatically, the absolute numbers of power tiller in use remains small in relation to the total number of farm households. There are nearly two millions two-wheel tractors in use, that is to say an average of one tractor per 58 rai of arable land. In addition, the use of these small tractors is heavily concentrated in Central Plain.

It is interesting to compare the growth rate of power tillers to the one of draft animals. The latter was constant until the end of the sixties at around 2% per year. After that, it declined rapidly to about 1% per year from 1970 to the mid 1980s. From the mid-1980s to now the absolute number of buffaloes declined by more than 2% per year. Nowadays animals are no longer used for cultivation in the irrigated villages.

The major change in the threshing method has been through the introduction of axial flow rice threshers which occurred in the mid-1970s. This rice thresher was first designed by the International Rice Research Institute (IRRI) and then locally produced. It was a remarkable success and has rapidly been adopted. The number of threshers produced increased from around 4,000 units in 1975 to nearly one million in 1999. Other machines such as sprayers and fertilizer throwers have also shown a steady increase. They were developed at the same time as pesticides and fertilizer use increased. Although fertilizer and pesticides can be applied by hand (or hand pump), the use of machines achieve higher precision and reduce waste.

Finally, it is worth underlining that government policy towards mechanization has been of minor importance. Machine prices have not been significantly affected by market intervention. Prior to 1982, taxes and tariffs charged on the import of complete sets of machines were lower than those on local products and in 1982, the local machinery producers were successful in pressing the government for increased protection (Onchan, 1985). There seems to have been little government intervention too in the establishment of the agricultural machinery manufacturing industry, although The Engineering Division of the Department of Agriculture has long been involved in developing farm machinery adapted to the Thai environment. In fact, public sector research has contributed little to machinery development due to the relatively high costs of the machines developed compared to imported or local machines developed by private firms. These firms have undertaken most of the research and development of farm machinery; in many cases, its application and adaptation were carried out almost exclusively by small manufacturers, working close to customers and often at customers' request.

3.2.2 Patterns of Agricultural Equipment among Rice Farmers

Table 3.9 shows the distribution of agricultural equipment owned and used in the sampled areas expressed as the percentage of the number of holdings reporting the ownership of farm machines to the total number of holdings of each size class. Two wheeled tractors are the dominant type of machinery, and are well established in the irrigated areas. Ownership of other machinery, especially threshers and harvesters is very limited. Animal power has been almost totally displaced from rice production. Work animals are mostly confined to non-irrigated areas in Roi Et and Pitchit and can no longer be found in the irrigated areas of Suphan Buri and Pitchit which are more heavily mechanized. The extent of farm mechanization varies widely from one province to another. Farmers in Suphan Buri have tended to adopt most types of available machinery. In this area, mechanical technologies have supplemented the biological ones. Two-wheeled tractors appear to be owned by 84% of holdings, 64% of the farmers reported the ownership of a water pump and 31% reported the ownership of a sprayer. In Pitchit tractors were found to be heavily concentrated in irrigated areas where nearly 88% of the farmers own one or more tractors, and around 46% own a water pump. In rainfed areas, a few tractors are found, but almost 25% of farms own a water pump. The Northeast has a very small share of the farm machinery compared with the North and the Central Plain, the percentage of households reporting the ownership of tractors in place of animals is still relatively low. Less than 10% of farmers own a tractor whereas almost 85% of them own one or more buffaloes.

Table 3.9: Distribution of Agricultural Equipment Owned and Used

	< 15 rai		16-30 rai		31-50 rai		51-100 rai		101+ rai		All sizes	
	Own	Use	Own	Use	Own	Use	Own	Use	Own	Use	Own	Use
Suphan Buri												
Tractor	53.1	68.8	97.6	100.0	100.0	100.0	100.0	100.0	100.0	100.0	84.0	90.0
Water pump	53.1		61.9		77.3		100.0		100.0		64.0	
Thresher		53.1	2.4	40.5	13.6	50.0		66.7			4.1	47.0
Harvester		15.6		16.7		4.5		33.3		100.0		15.0
Transportation	3.1	37.5		52.4	4.5	36.4	33.3	66.7		100.0	3.1	45.0
Sprayer	15.6		31.0		50.0		66.7				31.0	
Pitchit irrigated												
Tractor	78.6	78.6	85.7	90.5	90.5	100.0	100.0	100.0	100.0	100.0	87.8	90.5
Water pump	35.7		33.3		42.9		64.3		100.0		45.9	
Thresher		37.5		19.0		61.9		28.6	50.0	75.0	2.7	39.2
Harvester				4.8				7.1				2.7
Transportation		21.4	4.8	23.8	4.8	19.0	14.3	21.4		25.0	5.4	21.6
Sprayer			4.8		14.3						5.4	
Buffaloes	7.1				14.3		7.1				6.8	
Pitchit rainfield												
Tractor		40.0	25.0	50.0	50.0	50.0	66.7	83.3	100.0	100.0	41.7	58.3
Water pump					25.0		66.7				25.0	
Thresher		40.0		25.0		12.5		16.7				20.8
Harvester		20.0										4.2

Transportation				25.0				16.7				8.3
Buffaloes			25.0		12.5						8.3	
Roi Et												
Tractor	3.3	6.7	9.1	18.2	25.0	45.0		100.0	100.0		9.7	17.5
Water pump			4.5		10.0						2.9	
Thresher		1.7		4.5		10.0						3.9
Transportation		6.7		18.2		20.0						11.7
Buffaloes	81.7		95.5		85.0						84.5	

SOURCE: ORSTOM/CUSRI Project, Survey on Rice Farmers

The benefits of mechanization are not evenly distributed among different groups of individuals within given areas. According to the figures in Table 3.9, there is a significant inverse relationship between the landholding size and the ownership of agricultural equipment. This comes both within regions where larger farms own equipment more often than smaller ones, as well as across regions where those regions with larger farm sizes have higher equipment rates than those with smaller farm sizes. For example, while all farms operating 50 rai and more have at least one tractor, 53% and 79% of farms, respectively, operating less than 15 rai own tractors in Suphan Buri and irrigated areas of Pitchit. As regards other farm machinery, the same pattern prevails: the ownership of water pumps, sprayers and means of transport is heavily biased in favor of large farms. In rainfed areas of Pitchit and in Roi Et, small farms depend entirely upon human and animal power and we can observe a significant inverse relation between buffalo ownership and the size of the holding in these areas.

Larger farmers are thus more dynamic in the adoption of technical innovations than their small counterparts. Multiple market imperfections which result in different factor prices for large and small farmers, are usually put forward to explain the fact that large farms adopt mechanical technologies faster than smaller ones. We have extensively discussed labor market imperfections in Chapter 2 and concluded that the rural labor market in Thailand function quite well.

As far as the price of capital is concerned, the group of large farmers is generally in a better position to obtain extension services, information and marketing arrangements at lower real costs than small farmers. Furthermore, they can better spread the fixed cost of a given practice over a larger area than small farms, thereby lowering average fixed costs. Large farmers are thus more readily able to purchase and utilize large units of capital as well as being particularly well-placed to take advantage of any capital-using innovations. Last but not least, the preferential access to institutional credit on the part of the larger farmers has permitted them to purchase mechanized inputs at a lower cost than small farms. These last tend to lack access to cheap and long-term institutional credit because they possess less collateral and are generally associated with a higher probability of default (Cf. Chapter 4).

There is a body of evidence showing that tenure normally affect the demand for inputs such as farm machines. Much of the literature argues that tenancy has a negative impact upon the propensity to invest because tenancy fails to provide farmers with both adequate incentives and means to adopt new technologies that could enhance production efficiency. Uncertainty may prevail in tenants' minds regarding the possibility of a reallocation by the landlord of the land covered by the current lease contract. Such a dependence generates uncertainty regarding the future size of the farm and thus uncertainty regarding the usefulness of capital. If some of the land is taken away, farmers left with excess capital may suffer losses even if the capital is mobile because of the transaction costs of liquidating capital.

Another claimed dynamic benefit of land ownership is through credit allocation. Tenant farmers have few assets to offer as collateral to both institutional and non-institutional lenders. Unless they buildup savings from their surpluses, they are not able to find investment funds enabling them to develop their business.

The relation between machinery ownership and tenure form is empirically much less clear than the relation between machinery ownership and farm size. On the whole, there is no clear tendency for machinery owners to be dependent on tenurial status. Although tenants are expected to own few machines, if any, this is not what the results from Table 3.10 show, except in rainfed areas of Pitchit and in Roi Et where tenants seem less inclined to hold farm machines. In irrigated areas, the proportion of farms owning water pumps and tractors is lower in the group of owner cultivators than in the group of tenants. In these areas, rice cultivation is performed under quite good environmental conditions and generate higher and more secure incomes than in rainfed areas. The resulting lower risk of default has induced institutions in charge of agricultural credit to channel funds to tenant farmers through credit groups or credit cooperatives (cf. Chapter 4). As a result, farmers dependence on land collateral for access to institutional credit has been far less important than in rainfed areas. In this context, the supply-side link between credit access, tenure and investment does not appear relevant.

Table 3.10: Distribution of Agricultural Equipment Owned and Used by Tenure

	Landlords owners		Owners		Part owners Part Renters		Tenants		All Groups	
	Own	Use	Own	Use	Own	Use	Own	Use	Own	Use
Suphan Buri										
Tractor	100.0	100.0	75.7	81.1	92.0	96.0	80.0	92.0	84.0	90.0
Water pump	69.2		59.5		64.0		68.0		64.0	
Thresher	15.4	46.2	2.7	45.9	4.0	64.0		32.0	4.0	47.0
Harvester		23.1		13.5		12.0		16.0		15.0
Transportation		30.8	8.1	32.4		52.0		64.0	3.0	45.0
Sprayer	38.5		35.1		32.0		20.0		31.0	
Pitchit irrigated										
Tractor	100.0	100.0	87.0	87.0	87.5	93.8	87.5	100.0	87.8	90.5
Water pump	25.0		43.5		50.0		62.5		45.9	
Thresher		25.0	4.3	47.8		31.3		12.5	2.7	39.2
Harvester						6.3		12.5		2.7
Transportation			4.3	26.1	12.5	18.8		12.5	5.4	21.6
Sprayer			4.3		12.5				5.4	
Buffaloes	25.0		4.3		6.3		12.5		6.8	
Pitchit rainfield										
Tractor			36.4	63.6	83.3	83.3	16.7	33.3	41.7	58.3
Water pump			45.5		16.7				25.0	
Thresher				18.2		33.3		16.7		20.8

Harvester								16.7		4.2
Transportation		100.0				16.7				8.3
Buffaloes			9.1				16.7		8.3	
Roi Et										
Tractor	22.2	44.4	8.0	14.7	13.3	13.3		25.0	9.7	17.5
Water pump	11.1		2.7						2.9	
Thresher				4.0				25.0		3.9
Transportation		11.1		9.3		26.7				11.7
Buffaloes	88.9		84.0		93.3		50.0		84.5	

SOURCE: ORSTOM/CUSRI Project, Survey on Rice Farmers

Machinery is frequently hired and is therefore much more widely used than the ownership figures suggest. For certain operations, mechanization has widely spread to small and/or tenanted farms when machinery has been made available on a lease basis. Hence, the number of machines reported as being used on holdings is quite different from the total number of machines in existence.

Rental markets for land preparation with the use of tractors is now common. In Suphan Buri and irrigated areas of Pitchit the proportion of farmers using tractors is not very different than the proportion owning them, except for the smallest farm sizes. In the largest size groups, the ownership of tractors is already quite widespread. In contrast, in Roi Et where so few farmers own agricultural equipment, tractors have been leased or borrowed by a large number of farmers and while endowment ratios differ among farmers' group, factor-use ratios appear very much closer.

Hiring also appears very important for thresher and transportation. The comparison between the distribution of the ownership of these machines and the distribution of the equipment used indicates that they are being made available on a contract basis to large and small farming units. The rental of rice threshers has been far greater in double-cropping areas, where threshers relieve the constraint on available labor at the time of harvest, while buffalo treading remains the common method used in single-cropping areas. Almost 50% of rice farmers in Suphan Buri, 40% in irrigated areas of Pitchit, 21% in rainfed areas use threshers, whereas less than 5% of them own these machines. Except in Suphan Buri where 15% of farms reported the use of harvesting machines, the harvesting operations are still mainly done by hand.

To summarize, the proportion of small farms and tenants using equipment such as tractors, threshers, harvesters and means of transport is much larger than the proportion owning them. However, despite higher utilization rates achieved by a well-developed system of contracting, lags in adoption rates owing to size or tenure still hold true, especially in rainfed areas. Therefore, although access to land ownership is certainly an important factor in encouraging investment, the imperfections in other markets and the corresponding biases in the incentive structure posed by current prices for inputs are also at work.

3.2.3 Explaining the Mechanization Process

3.2.3.1 Industrialization and Growing Labor Scarcity

In industrialized economies, agricultural mechanization has generally formed part of the response of farmers to the growing income gap between their incomes and those of industrial workers. In Europe and North America, mechanization has been raising farm output per person, and increasing the scale of cultivation has represented the key to this process. Thai farm households have responded to industrialization in a different way. Because there was both little scope for extending areas under cultivation and large opportunities of employment in the industrial sector, they have engaged in a strategy of multiactivity. To improve their livelihood, farmers have integrated industrial work as a source of income through off-farm activities while continuing to practice rice cultivation. Therefore, rather than enabling farmers to increase their scale of cultivation, machines have made it possible to perform the heavy tasks of rice cultivation for part-time cultivators.

The pattern and speed of mechanization in Thailand has been heavily influenced by the growing scarcity of labor and land. Under the conditions prevailing up to about the 1950s, rice cultivators had little interest in mechanization. The labor force engaged in rice farming was growing at around 2% per year, a rate much lower than the rate of growth of planted areas. There was no point in introducing a labor-saving technique in this system with large labor supplies. Throughout this period, availability of labor was the main factor limiting mechanization. From the 1950s to the 1980s, the industrial boom decisively shifted the center of gravity of the economy toward industry. Despite the continuous outflow of temporary workers in industrial activities, the growth of the labor force engaged in rice farming still grew rapidly, while the growth of cultivated paddy land decelerated from nearly 2.7% to 2.2%. Interestingly, the share of off-farm income in total farm income increased from almost zero to nearly 30% over this period. This phenomenon explains why mechanization occurred despite a declining land/labor ratio. The 1990s appear as a decade of increasing land/labor ratios because the number of rice workers is declining more rapidly than the area planted. This, in turn, supports a second movement toward rapid mechanization.

The expansion of machinery use thus occurred in Thailand when labor was moving rapidly off the land, that is in the period and areas where labor costs (relative to capital) were rising. It would be helpful to examine the degree to which rural wages have changed over time. Unfortunately, the examination of rural wage levels is hampered by the absence of comprehensive wage surveys and the consequent scarcity of detailed and reliable wage data. Only data on wages for urban unskilled private workers are available from Labor Force Surveys. Data reveal a clear increasing trend in the real wage of these workers over the period from the 1980s to now. It is a fair presumption that rural wages increased over time in the line of urban wages because there are strong labor market linkages on both the supply and the demand side among sectors. Over time, there has been a shift from agricultural own-account and unpaid family workers into urban wage workers as wage employment opportunities expand, thereby reducing labor available for rice farming. The fact that households members can shift into seasonal urban wage employment is another important factor that push agricultural wages up.

The rapid growth in ownership of agricultural machines has also been closely tied to the availability of institutional sources of credit. Since most investment outlays require a substantial amount of liquid resources, investment typically cannot be financed from one year's income; long or medium-term credit, or accumulated savings are needed. Credit therefore appears critical to the modernization of farms and there are strong reasons to expect machinery adoption to be conditional upon borrowing. Until 1975, the small supply of cheap institutional credit probably curbed the evolution of mechanization. Since 1975, large increases in credit on favorable terms have attempted to encourage adoption of new technology. This increased quantity of agricultural credit has permitted farmers to purchase more mechanized inputs which they could not provide from their own resources (cf. Chapter 4).

The regional differences in the relative scarcity of labor have been significant because various combinations of production techniques that affect the demand for labor are observed. Actually, the shortage of labor has been increasingly critical in the intensive irrigated rice farming areas of Central Plain. The greatly decreased flexibility in timing agricultural activities with double cropping as well as the increase in the number of labor days required by higher cropping intensity have certainly contributed to the increased difficulty in finding hired labor and to wage rate pressures in irrigated areas. By contrast, in rainfed areas the seasonality in labor demand would tend to suggest the existence of a substantial labor surplus, at least during the dry season. These differences in labor requirements between regions are consistent with the regional pattern of diffusion of mechanization. This suggests that shifts from animal to mechanical power have been induced, at the local level, by labor bottlenecks and related wage increases.

Another way to overcome local labor shortages could have been through migration, which can be considered as an alternative to mechanization. Regions where rapid agricultural development takes place could import labor from stagnating areas through seasonal or permanent migration, with subsistence farmers on marginal land constituting the incremental labor force migrating to more commercialized areas. Consistent with national data, our survey showed that intra-regional labor flows are by far more important than inter-regional migration (Chapter 2). The biggest source of migrant labor in any region is from that region itself. Migration over long distances is clearly less significant than that between provinces in the same region.

Many factors contributed to mitigate the role of inter-regional migration in the reduction of regional labor imbalances. First, for a very long time, land availability for settlers in the North and Northeast have limited the supply of labor to the Center in the rainy season. Up until the 1980s, it was rational for many peasants to migrate into the forest areas to open up new land and actually take possession of the land (often illegally). The migration to remote areas was a best choice for the rural labor force than the migration to already settled areas where people would have worked as agricultural laborers. Crop prices were high and rising, and the move also increased the asset position of migrants since they took possession of the land initially opened up.

Second, wages differentials are usually among the most significant explanatory variables in the migration decisions. The Thai economic boom of the 1980s led to the rapid development of Bangkok and of the provinces surrounding Bangkok. Thus, when additional labor had to be attracted in paddy fields of Central Plain from outside by wage rate rises, rice farmers had to face competition from the Bangkok manufacturing sector.

On the whole, the migration process did not ease the local labor bottlenecks, nor it helped in distributing some of the benefits of agricultural development from the richer dynamic regions to the poorer stagnating ones. The widespread use of machines in Central Plain appears to have emerged mainly in response to sharply increased labor demand which could not be met from the rest of the country.

3.2.3.2 The Timeliness Argument and Potential Yield Increases

The importance of relative factor scarcity and related factor prices does not mean that mechanization should not be observed in areas where land is scarce or population densities are low. According to Binswanger (1986), whatever the relative factor endowments, the machines could be profitably used for those operations where concentrated power and/or speed provide them with a comparative advantage relative to animal-drawn implements. One of the issues that has attracted great attention in the literature on machinery selection has been the impact of machinery use on the timeliness of agricultural operations.

Land preparation is the heaviest task of the agricultural year and requires male labor, with or without a draft animal. How advantageous the tractor is compared to the buffalo for land preparation depends mainly on land expansion possibilities, cropping intensities and the time available for land preparation. Land preparation can be a major obstacle with regard to the introduction of double cropping because double cropping shortens the period of time within which land must be prepared. In the irrigated areas, desirability of mechanization depends thus on whether or not the potentials of double cropping can be realized without tractors. If no, a major argument in favor of mechanization of plowing is that it allows faster land preparation.

Across provinces the relation between the type of power and cropping intensity is as one would expect: buffalo owners of Roi Et have the lowest cropping intensity while tractor owners of Suphan Buri and Pitchit experience higher cropping intensities. This might tend to provide support for the thesis that tractorization is an important factor contributing to an increase in cropping intensities. However, in Thailand, government investment in irrigation projects as well as farmer investment in water pumps had already increased cropping intensity as a result of better water supply and control before the introduction of the tractor. Both past and present agricultural systems have thus achieved double cropping without tractor use. In rainfed areas, the main bottleneck is the lack of opportunity for dry season cropping due to lack of irrigation facilities for that season. At that point tractors will not lead to any increase in multiple cropping.

In addition, within the same province, the greatest intensity of cultivation occurs in the smallest farm size despite the fact that large farms are usually in a better power position. The survey data report that the multiple cropping index for holdings less than 15 rai in Suphan Buri is 172, whereas there are only 53% of these farms which own power tillers and 69% use them. Conversely, all holdings of more than 30 rai own and use tractors with higher occurrence than small farms, although cropping patterns are much less intensive. We must therefore conclude that tractors might not have been a significant factor in intensification in double cropped areas.

Another rationale for introduction of machines is that in many cropping systems, delays in sowing or harvesting operations can greatly reduce crop yields. The introduction of new sources of power which make more timely planting possible, could be beneficial even in non-irrigated areas. In dryland agriculture crop yields are typically a function of the day of sowing. Plowing must be done at the right time and the suitable period is limited by rainfall: immediately after pre-monsoon rains and before the land is flooded by the monsoon rains. Investing in power tillers enable the crop to be sown sooner, resulting in higher yields, but at a cost of higher investment in machinery.

It is difficult to test from the survey data whether farmers who mechanized some operations exhibit higher yields than those who did not. Mechanization has usually been associated with other improved practices such as double cropping, fertilization, chemical weeding which also have strong impact on yields. However, empirical evidence gathered tend to suggests that farmers who mechanize do so mainly to save time and/or labor, not to raise yields.

3.2.3.3 Specific Operations Costs

Farmers change their techniques only if they think that the changes will reduce the unit cost of production. When the opportunity cost of labor and/or the cost of maintaining bullocks becomes sufficiently high, it makes sense to shift to other power sources. If we assume that all agricultural operations for a given crop production process can be performed either by using labor (family, hired or a combination of the two) and buffaloes or by using machines, the farmer can choose an investment pattern or a factor combination such that each operation is performed by that technique which has the lowest cost.

It is therefore interesting to explain in terms of relative costs why particular operations attract investment first while traditional techniques still prevail for other operations. We concentrate here on the two operations that are potentially machinery-using, namely land preparation and harvesting. Unfortunately, the cost of land preparation according to different levels of mechanization (animal combined with labor or power tiller combined with labor) has not been calculated because we lack data for such an estimation. Brief remarks will first be made on the economy of tractor ploughing in comparison with buffalo ploughing. Then for harvesting operations we proceed to develop the analysis by comparing hired labor costs and costs of renting machines per rai.

The main advantage of tractors for land preparation is that a tractor can in principle work 24 hours a day at the same level of efficiency. Buffaloes are incapable of working a higher than normal number of hours when the workload increases so that higher cropping intensities can only be achieved by increasing the number of animals. Since the cost of a work animal is practically fixed, this leads to high buffalo costs per hour and to a non-competitive position vis a vis the tractor. In addition, where there is abundant grazing on pastures and grass fallow, the cost of keeping cattle is low, because farmers leave their harvested fields to be grazed by any animals. In contrast, where grazing land is scarcer or has disappeared, or where the land used for the cultivation of fodder crops has alternative, more profitable uses, farmers may find it cost-effective to mechanize operations previously animal-powered. In double cropped areas it is likely that relative prices work in favor of mechanization of ploughing.

Harvesting and threshing are often performed on a custom basis; in such a case they represent a variable rather than a fixed cost to farmers and can thus be compared to the cost of hiring labor. However, this comparison is based on strong hypothesis. First, animals and machine are perfect substitutes of labor, not complements. Actually, animals and machines do not perfectly substitute for labor since they require at least one man for operation, while the cost of renting a machine often includes a wage premium for the operator. Second, we implicitly assume that the amount of family labor is left unchanged whether machines or hired labor are used. Although labor supervision is probably of minor importance among rice farmers due to a low average holding size, it remains that mechanization of agricultural operations not only has the effect of replacing workers but also of disciplining their efforts. Thus labor supervision on the part of family members is certainly less intensive when machines rather than hired labor are used. Third, the cost of renting machines is assumed to reflect a competitive market for contractor services and to be close to the opportunity cost of capital.

Table 3.11: Labor Compared To Thresher and Harvester Renting Cost Per Rai (in baht)

	< 15 rai	16-30 rai	31-50 rai	51-100 rai	101 + rai	All sizes
Suphan Buri						
Hired labor cost	344	369	340	347		353
Renting machines cost	264	330	200	160	300	283
Pitchit						
Hired labor cost	161	211	228	192	199	202
Renting machines cost	170	160		180		170
Roi Et						
Hired labor cost	129	168	120		257	140
Renting machines cost						

SOURCE: ORSTOM/CUSRI Project, Survey on Rice Farmers.

We compare in Table 3.11 the cost of hiring workers with the cost of renting machines for harvesting operations. Although these figures should be considered with caution, they are instructive in interpreting the patterns in machinery use. They provide evidence of cost advantage of more capital-intensive techniques for harvesting operations in areas of the country where labor markets have been tight. In Suphan Buri and Pitchit, labor costs per rai for harvesting and threshing have advanced more rapidly than the cost of renting respectively a harvester and a thresher. In these two provinces the combination of hand and machine harvesting significantly reduce harvest costs relative to pure hand-harvesting. This gap between the two techniques is mainly due to the fact that rice harvesters and threshers are used in the period of peak demand for labor in crop production during which agricultural wages reach their maximum. By contrast, in Roi Et the use labor is much cheaper than in the other two provinces and the renting of machines during harvest time is rarely observed.

3.3 Conclusion

For almost one century, there had been little change in the way Thai rice was grown throughout the Kingdom. Expansion in planted area through forest conversion and increased cropping intensity of the existing paddy land, coupled with a fast population growth accounted for most of the growth in Thai rice output. Increases in land and/or labor productivity were of minor importance. Rice yields per hectare of land remained largely unchanged. Since the 1980s Thailand is experiencing a change in its relative resource endowment so that land, labor and water for rice farming have become increasingly scarce. Arable land suitable for rice cultivation has already been exploited to its maximum, the non-agricultural sector began to absorb workers at a faster rate, and rapid urbanization diverted more water to cities. All of this resulted in a considerable slowing down of rice ouptut growth.

Despite substantial public investment in irrigation for rice farming from the 1950s to the 1980s, Thai rice cultivation has remained dominated by rainfed farming. Today, a small proportion of the irrigated area is capable of yielding two crops a year because of a lack of water in the dry season. Even in the wet season the irrigation schemes do not guarantee sufficient water since the ultimate source is rainfall. Moreover, the water stored is not being utilized efficiently because of inadequate construction of terminal networks, and poor maintenance of the existing network. Figures have been mentioned about existing irrigation systems working at only 25-50% of their potential, a phenomenon related directly to the above reasons. Therefore in many areas, inadequate or unreliable sources of water are a serious obstacle to increasing land productivity.

Thailand is among the countries with the lowest use of the green revolution package in Asia, yet it is the biggest rice exporter in the world. The proportion of total rice area planted to HYVs remains circumscribed to small areas. Poor water control facilities have led to continuing low levels of adoption of Green Revolution technologies. Behind this poor result there is also the relatively profitable sub-segments of both international and local markets for traditional, low-yielding but highly tasty paddy varieties which command high prices induced by consumer taste preferences. The predominance of traditional varieties in Thai paddy fields explains, in turn, why average fertilizer consumption has remained low. Price incentives to use fertilizer have also long been low owing to the price depressing effects of taxation on rice. The recent devaluation of the bath which led to prohibitive prices of fertilizer, contributed to further reduce fertilizer purchases among rice farmers compared to previous years.

Beginning in the 1950s, mechanization in Thailand has tended to follow a sequence starting with the use of motors for static tasks such as raising water from canals. Later on, the mechanization extended to power-intensive operation such as land preparation. The use of two-wheeled tractors became widespread during the 1970s and has then been on an upward trend. Mechanization subsequently developed with the adoption of sprayers, threshers and harvesters. However, capital investment in machinery other than tractors remains limited.

The mechanization process has been basically dominated by a decrease in the rural labor force and rising rural wage rates that were associated with a growing demand for labor in the urban industrial sector. Locally, the mechanization process has been reinforced by the growing labor scarcity resulting from increases in labor demand arising out of expansion of double cropping.

On the whole, technological changes have been uneven. The advances made in certain areas contrast sharply with other areas and regions where farming is still carried out by traditional methods, being entirely dependent on the monsoon rains and the power of draft animals. Most capital investments have been confined to the more prosperous areas where there has been a substantial investment in physical and institutional infrastructure development. The pace of agricultural change has been extensive in the irrigated areas of Central Plain which have enjoyed a greater rate of growth of production and incomes. The North is somewhat behind and still in the process of transition from traditional to new technologies, whereas the Northeastern region, with natural constraints based on soil types, has lagged considerably and is only in the early phases of adoption.

Although small farmers have little land and a larger supply of family labor while large farmers have better access to credit and better managerial ability, both large and small farmers invest in current inputs such as fertilizer, herbicides and pesticides. Actually, small farms use intermediate inputs (fertilizer, herbicide, pesticides) and irrigation far more extensively than do larger farms. In contrast, the possession of agricultural equipment is strongly correlated with the size of landownership. Tenurial status was a relatively significant determinant of rice technology in rainfed areas but not in irrigated ones as a result of good access to institutional credit. Factor-use ratios appeared more independent of size and tenurial status in rainfed areas due to the existence of rental markets for machines.

Chapter 4

ACCESS TO RURAL CREDIT

4.0 INTRODUCTION

Imperfect credit markets prevailing in developing countries have often been considered as preventing agriculture from expanding. Difficulties in access to adequate financial services coupled with high interest rates charged by private moneylenders are believed to affect adversely the utilization of modern inputs, dictate factor ratios that deviate from optimum, and ultimately lead to lower output. The definition of agricultural credit policies in developing countries has generally reflected the concern of policy makers that a shortage of institutional credit constrained growth in this sector and prevented the farmers from adoption of new technologies. Strongly convinced that the private moneylenders possessed considerable market power, most governments have thought that this power should be curbed by competition from institutions. The policy response arising from this vision has been to provide to farmers large amounts of cheap institutional credit. This orientation appeared doubly attractive, promising both to lessen the farmers' dependence on, and exploitation by the money lenders and to stimulate the process of agricultural growth.

However, directed credit programs have been the subject matter of considerable debate since the 1990s. Critics increasingly argued that the provision of cheap credit at a low, disequilibrium interest rate led to a large number of unsatisfied borrowers and that many of these programs resulted in huge costs, with little identifiable impact on farm production and investment.

For about a century, Thailand has followed a deliberate policy to build and strengthen formal credit institutions with the intention of supplanting the village money lenders. Working on the assumption that there was a large and unmet demand for institutional farm credit, the government established specialized credit institutions to channel increased amounts of credit to rural areas. However, prior to 1975, formal credit institutions supplied insignificant amounts of loans to the rural sector. Dramatic changes occurred in the agricultural credit policy in 1975 and the structure of the rural financial markets was transformed. That year the Bank of Thailand ordered the BAAC as well as commercial banks to expand their loan portfolio to farmers so as to adequately serve the credit needs of all segments of farmers' population. The government also authorized the Bank of Thailand to enforce the commercial banks to play a greater role in agricultural lending. Since then, the Bank of Thailand has prescribed increasing agricultural credit targets for the commercial banks. An initial target of 5% of the total stock of loans and advances outstanding was set at the end of 1975, raised to 7% in 1976 and 9% in 1977. By 1978 this rate had been stabilized at 11% of total deposits. At the same time, the Bank of Thailand relaxed the control over opening new commercial bank branches. Opening new branches in up-country locations was allowed provided that the branch guaranteed to extend to farmers 20% of its total deposits. This policy resulted in a massive increase in the flow of credit directed to the agricultural sector from less than 5 billion baht in1975 to nearly 351 billion in 2000.

In spite of these efforts, it is widely held that there exists a substantial gap between the demand for and the supply of agricultural credit in Thai rural areas; that the institutional sector, in particular, have been unsuccessful in the supply of credit services to marginal farmers; that preferential access to formal credit in favor of better land-endowed farms located in the more productive regions have exacerbated existing levels of income inequality. In addition, optimism concerning the eviction of rural private lenders by public credit agencies has been opened to some reservations. Despite the competition institutional lenders have provided, private moneylenders have not been completely displaced from the credit market. Finally, it is increasingly held that policy interventions in other markets could increase the call on credit to a greater extent than credit market interventions.

Using the surveys results, this chapter presents a detailed analysis of the distribution of both institutional and non-institutional credit among rice farmers and attempts to evaluate its effects on the adoption of technical innovation. Section one provides an overview of the various sources of credit available to rice farmers, the spatial distribution of credit flows, the characteristics of the loans farmers obtained. Section two discusses the factors affecting the demand for agricultural credit and investigates the characteristics of farmers who had access to credit in terms of their farm size and security in land ownership. Section three tests the assumption that credit shortage limits the use of variable factors and constrains investment in fixed capital.

4.1 GENERAL STRUCTURE OF CREDIT MARKETS

Rice farmers' credit needs are served by two distinct groups of lenders. The first group includes the principal institutions involved in mobilizing and allocating resources for lending to agriculture; they are the Bank for Agriculture and Agricultural Cooperatives (BAAC), commercial banks, farmers' cooperatives and farmers' associations. The second group of lenders covers a wide range of diverse individuals such as friends, merchants, relatives, neighbors, professional private moneylenders who lend money on a permanent or occasional basis.

4.1.1 Institutional and Non-Institutional Lenders: Market Share, Loan Length, Loan Size

The first public agricultural credit agency, the Credit Co-operative, was created in 1916 with the aim of giving loans to its members for agricultural production. A shortage of funds induced the government to create a new central agricultural credit agency in 1947, the Bank for Co-operatives. It was dissolved in 1966 and replaced by the Bank for Agriculture and Agricultural Cooperatives (BAAC), which has been in operation up until the present time. The BAAC was established as a public bank responsible for the provision of credit to farmers for agricultural activities[33]. It is the only governmental agricultural credit organization.

[33] Since 1999, the state-run bank has been legally empowered to extend credit to farmers for activities outside the farming sector. In the past, efforts to amend the laws governing the

Commercial banks began to extend credit to farmers in 1961, but they were marginally involved in rural lending at that time. Prior to 1975, only four[34] on the twenty nine commercial banks in Thailand did much farm lending and the amount of credit offered to farmers was less than 2% of their total credit (Meyer, 1979). After 1975, commercial banks were able to comply with their agricultural lending obligations either through their branch offices directly with farmers, or they could to deposit the target amount at the BAAC. In practice, commercial banks met a large part of their lending requirements through substantial deposits at the BAAC because they had neither the network nor the administrative capacity to extend a large number of new loans to farmers.

Cooperatives and farmers' associations are the nation's third largest agricultural lender.[35] Cooperatives and farmers' association have a long history in Thailand's agricultural credit system. Since its inception, the cooperative movement has been a government-sponsored institution. Founded in 1916 with the aim of freeing farmers from reputed heavy debts and exorbitant interest rates, the cooperative movement made little progress until 1932. The Cooperative Society Act of 1928 and the Civil Association Act of 1940 provided legal status for the further development of various types of cooperative. By 1951, it was estimated that 5% of all farmers were members of some type of cooperatives (Ingram, 1950). According to the Agricultural Extension Department, more than three thousand agricultural cooperatives were registered in 1999, bringing together nearly five million members; and there were 3,964 farmer groups totaling 514,188 members across the country.

Cooperatives and farmers' associations acquire funds from several sources. Members are an important source of funds through deposits and ownership capital, but most cooperatives' liabilities are supplied by the BAAC or commercial banks. The credit function is generally used for purchasing equipment and for long-term investments in fixed assets, such as irrigation and other land improvements, and for relending to farmers. Some cooperatives add

---

BAAC had failed because of fears the bank would deviate away from its prime objective of supporting the agricultural sector.

[34] They include the Bangkok Bank, Krung Thai Bank, the Thai Farmer's Bank and the Bank of Ayudya.

[35] The survey did not provide details on the type of association farmers are involved with. Cooperatives and farmers' groups are thus treated together.

the sale of agricultural inputs to farmers, usually through a system of an open account with a monthly interest charge.

The last group of lenders, generally called informal lenders, is the fourth source of credit. This group is subject to relatively little regulation and the lender itself rather than depositors is ultimately responsible for the obligations of its activity. The role of moneylenders has been of great importance in the past as a source of credit, when institutional credit organizations were less developed. Time series data on informal credit do not exist. According to various sources (Siamwalla, 1990; Onchan, 1985), the market share of informal lenders has declined in every region of the country as a result of the increase in the market presence of institutional lenders. In 1960, it was estimated that formal credit, as a percentage of total agricultural credit, was only 5% and by 1975, the percentage of formal credit had risen to 57%. Nowadays, it is widely held that the institutional sector holds more than 80% of farmers' debts. Therefore, conventional view that private moneylenders hold most of the agricultural debt is no longer acceptable. Competing supplies of institutional credit as well as regular attempts from the BAAC to help free farmers from the grip of the informal money market appear to have greatly eroded the moneylender's position.[36]

An interesting feature of the private credit market is that, as is typical in many developing countries, it operates on the basis of personal relationships. The borrower usually knows the lender either as a commercial partner or through family and neighborhood connections. The informal credit market thus appears segmented, with transactions taking place within small circles of related households. In the less commercialized areas like Roi Et, most informal contracts were provided by relatives, while in more commercialized areas traders[37] predominate as a source of informal credit: they managed nearly 40% of the informal loans in Suphan Buri and Pitchit.

[36] For example, in 1995 and 1997, the government allocated specific funds (two billion baht each year) to the BAAC in order to help farmers to clear debts incurred with informal lenders. Farmers who had borrowed from or mortgaged their land with informal lenders could apply to the BAAC for specific loans and use the proceeds to redeem the mortgaged land.

[37] The survey did not separate explicitly grain traders from village shopkeepers, whose reasons for advancing loans to farmers might be very different. However, a survey on rice marketing was also conducted by the author between July 1992 and July 1994 on a sample of 75 rice mills and 30 rice traders. Only 4 mills reported credit transactions and traders mentioned here

The distribution of loans by source drawn from our survey is shown in Table 4.1.[38] At least 65% of the borrowing contracts were issued by the institutional sector. The BAAC is obviously the leading source of institutional credit for rice farmers: the number of BAAC contracts accounted for one third to one half of each province's total contracts. Table 4.1 also indicates the importance of cooperatives and farmers' groups as a source of credit, keeping in mind that the sources of cooperatives' funds are dominated by the BAAC. The number of contracts signed with cooperatives is almost equal to the number signed with the BAAC in Suphan Buri and Roi Et, but lags far behind in Pitchit. Commercial banks, in comparison, were the source on average of less than 6% of each province's total number of loans. These results tend to indicate that the supply leading approach has been successful in reaching farmers that would not otherwise have access to the formal financial system.

Although the availability of services from the cooperatives and farmers' associations have been undoubtedly important in reducing the reliance of their members on private lenders, the agreement concerning the eviction of private lenders from the rural credit system is open to some reservations. This source of credit still remains quite significant. Not only farmers in the poor region of the Northeast depend on informal lenders. The share of non-institutional loans in the total number of loans ranges from roughly 34% in Suphan Buri to 25% in Pitchit and 22% in Roi Et. In the light of these figures, it can be concluded that the erosion of the informal lender's relative position has been probably much less marked than commonly claimed.[39]

---

are not rice traders. The grain market and the credit market do not appear interlinked, that is, rice traders and millers are generally not involved in credit transactions with rice farmers.

38 During the survey, farmers were asked about the amount borrowed from external sources, both formal and informal. All the lending contracts concluded and not closed at the time of the survey were registered.

39 Moreover, the data certainly underestimates informal credit because respondents were generally more reluctant to reveal what they had borrowed from non-institutional than from institutional lenders.

Table 4.1: Distribution of Loans Characteristics by Source

	Number of contracts	Percent of contracts	Size per transaction (000 baht)		Percent of short term loans	Average interest rate	
			Short-term	Long-term		Short-term	Long-term
Suphan Buri							
BACC	37	34.6	15742	9667	91.7	13.2	12.6
Commercial banks	6	5.6	-	48000			11.5
Cooperatives	28	26.2	15833	39167	50.0	12.7	11.9
Total institutional	***71***	***66.4***	***16143***	***38895***	***69.7***	***12.8***	***11.9***
Traders and millers	14	13.1	17000	18500	72.7	31.6	26.8
Neighbors	10	9.3	10000		100.0	42.6	42.6
Relatives	12	11.2	14000	10000	40.0	37.3	34.7
Total non-institutional	***36***	***33.6***	***17417***	***16750***	***66.7***	***36.2***	***36.3***
Pitchit							
BACC	48	55.2	24406	35843	66.7	13.2	11.9
Commercial banks	5	5.7	-	121250			15.5
Cooperatives	12	13.8	19978	20000	90.0	12.9	12.5
Total institutional	***65***	***74.7***	***12434***	***51357***	***66.7***	***13.0***	***12.5***
Traders and millers	9	10.3	9250	17000	80.0	45.7	51.4
Neighbors	6	6.9	31000		100.0	45.0	
Relatives	7	8.0	13333	17500	60.0		42.6
Total non-institutional	***22***	***25.3***	***17000***	***17250***	***76.9***	***45.3***	***46.1***

Roi Et							
BACC	21	36.2	14111	37000	82.6	12.8	11.5
Commercial banks	2	3.4	-	20000			18.3
Cooperatives	22	37.9	18450	33500	45.5	20.7	14.4
Total institutional	***45***	***77.6***	***15607***	***32778***	***61.7***	***14.3***	***14.2***
Traders and millers	2	3.4	10000	30000	50.0	42.6	19.6
Neighbors	3	5.2	3000		100.0	53.2	
Relatives	8	13.8	1000	1000	25.0	79.6	61.1
Total non-institutional	***13***	***22.4***	***4250***	***8250***	***50.0***	***57.2***	***55.4***

SOURCE: ORSTOM/CUSRI Project, Survey on Rice Farmers

It was difficult to clearly ascertain whether the loans from the data set are short-term or long-term. Farmers need credit for four general purposes: working capital, marketing, consumption, and long-term investment. The first three types of credit needs can be described as seasonal because they are due to the sharp fluctuations in the farmers' liquid asset position at different times of the year. Generally, the loans granted for such purposes have repayment periods of less than one year and can be considered as short-term credit. Credit is also sought to finance high cost farm investments, and farmers are unwilling or unable to meet the full cost of such expenses from their yearly own incomes. Long-term credit is therefore assumed to have repayment period of more than one year.

However, the universal prevalence of exogenous shocks in agriculture calls for consensual contract flexibility. Many transactions with non-institutional lenders have open-ended repayment, i.e., no explicit time limit for repayment, whatever the purpose of the loan. In this case, the contract is deliberately incomplete and the possibility of renegotiating the terms of the contract due to unpredictable cash flow variations is implicitly present. Such loans are often short-term loans rolled over from time to time, depending on the borrowers ability to repay. Even when the loan falls due, institutional as well as non-institutional lenders may prefer extend the duration of short-term loans if clients encounter repayment difficulties, rather than foreclosing on a defaulting borrower. The nature of the dilemma faced by an unpaid creditor is well known: throwing good money on bad money may be the only way to get both back. Thus creditors can be dragged into a situation of providing "long-term" credit to a debtor. Hence, it is likely that the number of effective long-term loans, that is loans with initial repayment period of more than one year, is overestimated in Table 4.1.

The predominance of short-term credit is a common feature of rural credit markets in developing countries. Figures in Table 4.1 indicate that institutional as well as non-institutional credit for rice farmers has been heavily weighted toward short-term maturities: 60% to 70% of loans distributed by both groups of lenders are short-term loans. Among institutional lenders, long-term credit tends to come from commercial banks. In each province, all the sampled contracts distributed by commercial banks were long-term. Surprisingly, private lenders do not hold a higher share of short-term loans than formal lenders. There are two main reasons for this result. First, formal credit purposes such as residential construction or social events which require large cash outlays. The absence of institutional credit for these expenses which typically cannot be covered from the farmers' short-run cash flow led to the participation of informal sources in longer term credit. Second, the predictable higher rate of default on loans from informal lenders and, accordingly, the higher number of ambiguous long-term loans may be largely responsible for this.

The relatively minor role played by long-term credit in agricultural investment is confirmed by farmers' answers about the sources of finance for each item of equipment they own: 71.4% of the equipment purchases has been financed by own savings, 16.1% by loans from the banking sector, 5% by loans from cooperatives, 4.2% by loans from non-institutional lenders and 3.4% by loans from various sources. Hence, in contrast to working capital, personal funds are the main source of investment for fixed capital such as machinery, means of transportation and so on.

With such a proportion of farmers reportedly using their own savings to finance major items of farm expenditure, the question arises as to why farmers did not make use of low-cost institutional loans. Another major question is why lenders, and especially institutional ones such as BAAC, have provided so few funds for investment purposes. An important point in these questions is the matter of collateral required. Credit for investment involves longer-term, larger amounts of money, which in turn, requires more valuable collateral in terms of land, machinery, houses or vehicles. It is likely that farmers who do not have easy access to long-term formal credit because they lack the security required tend to borrow short-term loans, more easily available, in excess of working capital needs, to help finance their investments (cf. infra).

Average loan size for institutions' clients varies according to the lender and the terms of the loan. As expected, the amounts lent on a long-term basis by the institutional sector are generally much greater than the amounts lent on a short-term basis. Interestingly, the size of BAAC's loans tends to be much lower than that of any other source. According to Siamwalla (1989), this might be due to the pricing rule of the BAAC which has restricted the funding sources and maintained a lower size of operations than it would have if allowed to charge the full marginal cost. One of the responses of institutional lenders, when lower interest rates are imposed, consists in supplying a large number of loans to broad target groups, but with a much lower amount. Thus, considerations of administrative costs have led the BAAC to set an implicit maximum size for loans.

Informal loans are generally considered too small for the kind of investment deemed necessary and accused of providing an inadequate financial basis for the modernization of Thai agriculture. Data available from the survey indicate that although the size per transaction in the informal sector is on average much smaller than in the formal sector, substantial amounts of credit were supplied by informal lenders. Moreover, unlike prevailing opinion, farmers in the more advanced regions have been the principal recipient. The average amount of short-term informal loans ranged from 17,000 baht per borrower in Suphan Buri and Pitchit (while the corresponding figures for the institutional sector are, respectively, 16,000 baht and 23,400 baht), to 4,250 baht in Roi Et.

4.1.2 Formal versus Non-Formal Interest Rates

Generally speaking, interest rates on loans are determined by a combination of factors such as the source of the loan, the size of the loan and its maturity, the risk associated to the borrower, and the nature of the security he can offer. They also vary according to the purpose for which the loan is sought. Taking into account these factors, interest rates on agricultural loans must be high for four main reasons: most loans are small, incur high expenditure for lending service, the level of risk in the agricultural sector is high, and many borrowers cannot provide tangible assets as collateral.

There is in Thailand a great gap between the interest rates charged in the formal and informal credit markets. Our investigations established that informal lenders charge very high rates, 27% per year being the minimum found in Suphan Buri, whereas 40% to 60% elsewhere is the norm. Among private lenders, merchants lend money to farmers at rates substantially below those offered by other informal creditors. Although loans from relatives are usually expected to carry no interest, actually, there were very few cases of farmers borrowing from relatives without any interest. Loans secured from relatives appear to carry rates higher, in fact, than those charged by other creditors. In contrast, the institutional lenders' practice has been to require interest rates varying between 11.9% to 14.3% per year, within the range of government-prescribed interest rates for different categories of loans. Short-term loans are offered at an annual rate of interest close to 13%, while long-term loans are offered at rates comprised between 11.9% per year in Suphan Buri to 12.5% in Pitchit and 14.2% in Roi Et.

Whether the high informal interest rates reflect « *usurious* » rates or whether they reflect higher risks which money lenders have to bear remains a subject of controversy. Whichever explanation is correct, it remains that the comparison between the sectoral interest rates must be considered with caution. A first important issue is the lack of standardization of financial commodities that makes this comparison very difficult. Although the traditional neoclassical framework assumes that exchange mechanisms are free of cost, various transaction costs are associated to various sources of credit and tend to obscure the effective cost of a loan. The price of credit manifests itself in more than one dimension. At least three kinds of transaction costs might be involved. These include information costs incurred in identifying and screening different opportunities; application fees, service fees, gifts and bribes; but the largest and most important transaction costs might be the expenses resulting from contract negotiation costs which can involve repeated trips to town so as to complete loan procedures, receive and repay loans. The opportunity cost of work time lost could be substantial if loan transactions are concentrated in planting and harvesting periods.

It is not easy to accurately evaluate financial intermediation transaction costs due to the wide dispersion of elements included in these costs. This is a complex issue that requires a large amount of data which were not collected during the survey. However, it is possible to give a rough outline of the hierarchy of intermediation costs between the two sectors, using qualitative information. According to farmers' opinions, borrowing from non-institutional lenders requires obviously lower administrative and transportation costs because they offer simpler and more convenient procedures for their clients and because the borrowers and lenders usually live in the same village or district. Conversely, the bothersome aspects of officialdom associated to loan processing may serve as entry barriers for many farmers. Therefore, differences in transaction costs possibly yield a set of households for which informal sector costs are lower than the formal rate which is uniformly fixed for all households by the government.

A second reason why informal credit conditions should not be compared directly with those from formal sources, is that the risk taken is different. If we consider that the interest rate which an individual is « *willing* » to pay may act as a screening device (those who are « *willing* » to pay high interest rates, may on average, be worse risks), observed high interest rates are a reflection of the risk the lender incur. In the presence of a high risk, the interest rate an informal lender receives deviates from the formal interest rate because there is a possibility that a portion of the loan will not be repaid. Significant losses in the lender loan portfolio is an expected cost to incorporate in the calculation of the interest rate.

The high observed informal interest rates and their alleged adverse impact on farmers have been the subject of many debates. They are many who argue that peasants can hardly afford to borrow at such rates, given the low return of farming. If they do, the high cost of the loan rapidly depletes their available resources and undermines their ability to repay the loan. As a result, peasants can never draw the surplus back into production to improve their productivity. Undoubtedly, high interest rates put a serious strain on farmers incomes, but condemning the « *evil aspects* » of informal credit does not explain why so many peasants are still able and willing to pay such a high price for obtaining rural credit. Using informal credit, whatever the purpose, is obviously a better alternative to peasants in Thailand than not using it.

Informal lenders should be seen as filling a role of risk-takers which banks are presently unwilling or unable to play. It is widely held that the BAAC has been unable to lend to the agricultural sector the full amount required, both because of the great administrative cost and risk of default involved by an increase in loan transactions to marginal farmers, and a traditional strong political pressure to keep the BAAC's lending rates low (Siamwalla, 1989). The burden resulting from the compulsory low levels of interest rates have reduced the bank's incentive to lend to those borrowers who offer higher risks, leaving them to seek loans from the unorganized money market. Actually, many farmers who borrowed money from private lenders only were unable to borrow elsewhere since they had already been rejected by the institutional credit organizations.

4.1.3 The Question of Collateral

The viability of a financial institution depends on the cost of funds to the lender, controlling default rates, holding administrative costs to tolerable levels and providing services to borrowers that are competitive with those provided by other lenders. Among these challenging constraints, the most serious problem facing a lender is to know the borrower's intentions and ability to repay loans because the default rate has a multiplying effect: default results in the loss not only of the principal and interest but also in the costs of funds and administration committed to loans. As there is usually a serious lack of information for assessing the credit worthiness of the farmers, the potential for adverse selection arises. In addition, even if the characteristics of the borrower were fully known to the lender, in an uncertain environment, poor returns from the borrower's activities resulting from bad luck confront the lender with the problem of moral hazard. Finally, when the loan falls due, the lender must recover the principal and interest, and he faces a potential problem of enforcement.

The interest charged on a loan contract may influence the characteristics of those who apply for this loan because potential debtors differ on how likely they are to comply with a particular contract. However, lenders cannot rely on interest rates alone in a screening procedure to identify « *good* » borrowers. They all try to carefully screen the client's viability to ensure loan performance. Acquiring information regarding the borrowers is always costly for any lender, even though for informal lenders such costs are lower than for institutional lenders. For the former, information is a by-product of living near the borrower or being part of the same kinship or family group, while the later may find it virtually impossible to screen and monitor their borrowers properly.

Lenders therefore try to solve the problem of default as well as mix the various modes of enforcement through the requirement of collateral. The expectation is that a collateral requirement will filter out bad borrowers and attract those better positioned to assess and manage risk. A collateral requirement partly or fully shifts the risk of loss of the principal from the lender to the borrower. Collateral thus increases the expected return of the lender and reduces the expected return of the borrower.

All assets are not suitable as collateral. According to Binswanger (1986), they must satisfy two conditions. It must be easy for the lender to appropriate the asset in case of default (appropriability) and to ensure that the collateral cannot easily lose value (absence of risk). Among all the assets eligible for collateral, land is in great demand, provided that the owner has legal ownership rights on the land utilized as collateral.

When land is in short supply as collateral, lenders can move to collateral substitutes like credit groups or cooperatives jointly liable for each member's loan. The lender gains three possible advantages from a borrowing group: a reduction in administrative and screening costs, peer pressure and the collective threat of enforced sanctions to ensure repayment of joint liability loans. Credit groups induce borrowers to engage in a kind of self-selection wherein local knowledge about each others' assets and capabilities is used to sort out candidates. The credit group makes use of latent information on individual creditworthiness that are not captured in more traditional credit evaluation processes relying heavily on observed wealth and income factors, and leads to an informational efficiency gain (Hupi and Feder, 1990; Wenner, 1995). Credit groups can be seen as a means of transmitting information on borrower creditworthiness, and partly solve the asymmetric distribution of information on default risk, which, in turn, lower screening costs.

Group credit also improves enforcement capacity through the loss of borrowing opportunities. Since all members of the group will be excluded from future borrowing if the group defaults, members have an incentive to use moral persuasion, social ostracism, and mutual insurance to honor the external commitment to the lender. Another significant result might be for farmers involved in those groups to be identified as eligible for bank loans as individuals because of a history of successful repayment in an informal borrowing group.

This process, however, results in homogeneous farmers groups with respect to social and economic characteristics, as the group will tend to exclude farmers perceived as more risky than the mean of the group. This actually limits who is able to borrow through farmers groups.

Conversely, in moving from individuals to groups, lenders encounter risks that may cause low repayment rates. Weak management, committee members' lack of training and sometimes corruption might result in little encouragement or enforcement for loan discipline. However, late or non-repayment by borrowers might be due less to the consequence of opportunism and carelessness than of poverty and unanticipated income fluctuations. As the liquidity and income of individuals who live near each other and engage in similar economic activities are often highly correlated, group lending arrangements are subject to the danger of portfolio concentration.

Reputation is another form of social collateral that can guarantee contract performance for debtors who have demonstrated their ability and willingness to pay back. The main motivation behind contract performance is the desire to preserve long-term relationships and to maintain sources of funds supply. A gradual approach to screening relies on learning progressively about a client's qualities through repeated contracts. In such a way, economic agents learn about each other and revise their prior knowledge. They may come to trust each other.

Personal recommendation is a common form of reputation mechanisms. An implicit guarantee is often provided with the recommendation, in the sense that the guarantor is sure to be bothered by the creditor and asked to intervene should a problem arise. However, in spite of its sophistication, this screening method offer little protection against good payers who, due to unforeseen circumstances, become bad payers. Moreover, when interest in preserving the relationship is lost, so is enforcement.

Finally, the drawing up of a tied contract in which credit is given in association with another transaction can also be a powerful substitute to collateral. The interlinkage serve to reduce moral hazard of adverse selection by permitting the use of the contractual terms in one transaction to alter an agent's behavior in another. Product-credit market interlinkage is an example of disguised form of collateral. Farmers are lent money on the condition that they sell their produce to a trader at a pre-determined price. The forward sale of the standing crop can be interpreted as the pledging of this crops as collateral. However, this institution entails two main shortcomings. First, pre-determined prices are usually believed to be an underestimation of the value of the crop at harvest. Second, the credit farmers receive bind them to a particular trader and, as a rule, this means an obligation to sell only to that person. This closes the borrower's access to other lenders as well.

In Thailand, land collateral requirement is not always imposed on borrowers. Short-term loans from the BAAC, the leading source of farm credit, are dependent on the guarantee offered by the borrower: it can be joint liability, two guarantors, mortgaged assets other than land; farmers can also pledge paddy for loans[40]; loans dedicated to purchasing fertilizer must be only supported by evidence that inputs are actually purchased. Commercial banks usually require a more restricted set of guarantees. They lend to individual farmers on a short-term basis using mortgaged land as collateral or through credit groups.

A distinctive feature of the Thai agricultural credit has been heavy reliance on groups of farmers as the main short-term credit delivery system[41]. In spite of many difficulties, the BAAC has been successful in maintaining financial discipline and, with its group-lending method, has achieved a satisfactory repayment rate at a small cost (Siamwalla, 1990; Tohtong, 1988) even though the scale of the operation has expanded over time to cover more marginal areas and more risky farmers.[42]

Non-institutional lenders drew heavily on information obtained through personal, social and business relationships in order to pre-select clients, or they rely on recommendations of previous clients.

Long-term loans from nearly all sources, formal and informal, involve the use of land as collateral. For those loans standard practice is for the borrower to deposit his land title with the lender. This prevents the debtor from borrowing from another source or from selling the land which would leave him free to default on his loans.

Finally, the survey on rice marketing found that the intelinkage of loans with real sector transactions was an uncommon technique to ensure contract enforcement. Rice traders and rice mills surveyed did not report making advance price payment to farmers, nor have they contracted in any kind of credit with farmers. Rice traders and millers are engaged in competitive purchases of commodities, and they are rather poorly equipped to engage in money-lending. In addition, farmers appeared to be completely indifferent

[40] For example, in 1997 the BAAC was asked to accept pledges of up to 2.5 million tons of paddy from farmers.

[41] The condition to loan eligibility is that each member of a borrowing group agrees to assume liability for loans acquired for other members of the group. Each farmer's liability is ordinarily limited to the ratio of his loan to the total loan for which the group is jointly liable.

[42] According to BAAC's data, between 1996 and 1999, average repayment rates of 82% are recorded for jointly liable groups in contrast to an average rate of 66% for comparable loans with individual liability

about who their buyers are, which is inconsistent with the credit and marketing ties argument.

4.2 The Bias in Farmers Access to Credit

The need to borrow ordinarily depends upon the difference between income and outlays. Credit transactions are intertemporal transactions allowing farms to have their expense path to differ from that of their income path. If one's own adequate funds necessary to meet cultivation costs are not available in time, short-term borrowing is likely to occur. Likewise, long-term credit is necessary because the farmers' liquid assets or savings from their surpluses are unlikely to properly cover the desired investment. Therefore, one expects to find considerable credit needs in an agriculture with pronounced seasonality in production, or/and in the more productive regions undergoing technological transformation.

The borrowing behavior of farmers is also expected to vary for different classes of cultivators. Data on loans must be analyzed along with factors such as land tenure and landholding size since these variables have widely been confirmed to be good indicators of the socio-economic situation of farmers.

4.2.1 The Farm Location

Financial markets are expected to be affected by the rate of regional development on both the demand and supply side. On the demand side, much depends on the level of technology at which the farmers operate. Farmers who face different investment opportunities should tend to borrow different amounts; the higher the rate of return on capital, the larger the tendency employ capital in production and thereby the greater the propensity to borrow. Conversely, farmers using traditional technology under rainfed conditions might be unable to increase output through the use of extra inputs due to the limitations of the physical environment and high risks of natural disaster. It is unlikely that these farmers use their own or borrowed funds for farm improvement simply because there are no attractive investment opportunities or because returns cannot be guaranteed. On the supply side, lenders should choose to locate in areas where agricultural opportunities are better, risk is lower and hence, chances for loan recovery are higher.

There are marked differences between the provinces surveyed in the incidence of credit activities (cf Table 4.2). The amount of credit distributed tends to correspond to the agricultural conditions. The province of Suphan Buri is associated with larger borrowings, not only from institutions but also from informal lenders. The number of contracts has been substantially higher in this province (107) where farm technology is more advanced and farmers earn higher incomes than in Pitchit (87). A higher rate of technical change combined with a high degree of commercialization of agriculture has clearly increased the demand for funds. In contrast, only 58 contracts have been signed in Roi Et. The limited opportunities for commercialized farming in this rainfed province, as well as the slow development of modern agricultural techniques has limited the size of the market for loans. Therefore, while only 18% of Suphan Buri households did not borrow from any source, the share of non-borrowers was 24.5% in Pitchit and 42.3% in Roi Et.

A noticeable feature of the distribution of loans is the inter-provincial variation in the farmers' dependence on non-institutional credit. Opposed to conventional belief, Northeastern farmers appear to have a higher probability of gaining access to institutional credit compared to farmers in other regions. Nearly 74.6% of the respondents in Roi Et borrowed from institutional lenders, 71.6% in Pitchit and 56.1% only in Suphan Buri. Therefore, whereas most of the credit needs are fulfilled by institutional lenders in the poorer provinces, non-institutional lenders appear to play a greater role in more advanced provinces.

Another common observation is that farmers are active borrowers in both formal and informal credit markets. Within each province, the number of cases where there are multiple loan source appears to be significant. Thus even those farmers who could get credit from institutions still receive some credit from informal lenders. Of the rice farms who reported some borrowings, 19.5% in Suphan Buri borrowed from both the formal and informal sectors compared to 16.2% in Pitchit and 5.1% in Roi Et. Interestingly, it appears again that moneylenders have tended to concentrate their operations in dynamic areas where average income has risen.

Observed low levels of farmers' participation in the formal credit market in Suphan Buri tends to corroborate that there exists a persistent excess demand for formal credit in this province. The development in the theory of financial markets has pointed towards equilibrium credit-rationing as an inherent feature of financial markets (Stiglitz and Weiss, 1981; Hoff and Stiglitz, 1990). The nature of the asymmetric information in the market for loans implies that risk-adverse lenders seek to avoid adverse selection of borrowers or tries to reduce the incidence of potential default through use of quantity constraints. As a result, many borrowers are unable to satisfy their demand for credit at the prevailing interest rate. In Suphan Buri, it is likely that credit needs are much higher than in other provinces owing to higher expenses on inputs, whereas average loan amounts do not exceed those granted to other provinces. The formal sector provides only partial financing for projects, thereby forcing borrowers to resort to the informal sector for the remainder of the funds required.

Agricultural performance also influences the conditions of credit (cf. Table 4.3). Farmers residing in areas characterized by rapid growth in output and by the use of new technology appear to benefit from lower interest rates from both institutions and private lenders. In the relatively more prosperous province Suphan Buri, the average interest rate varied between 12.8% and 36.2% per annum, while in the relatively poor province Roi Et the average interest rates varied between 14.3% and 57.2% per annum. In addition, the gap between formal and informal interest rates is less wide in Suphan Buri and Pitchit than in Roi Et. This result contradicts the argument which states that where the share of institutional finance in total loans is high compared to the share of moneylenders, the gap between the rates at which credit is made available by the two sources should narrow. Actually, the gap between the two rates of interest dwindles in Suphan Buri where farmers rely more on non-institutional credit sources and tends to grow in Roi Et despite the fact that nearly 80% of the credit contracts come from institutional sources.

These results suggest that areas where productivity and incomes increased and stabilized have created a potential for a well-developed credit market, where both institutional and non-institutional interest rates are kept lower than in other regions. The availability of irrigation supported the adoption of more capital intensive techniques, which in turn, led to higher demand for cash. Better water control also reduced the probability of default due to adverse weather conditions and thus the risk premium attached to any loan, thereby lowering the cost of borrowing. An important policy implication is that interest rates can be lowered indirectly through the provision of technical change, investment opportunities, government expenditure on rural infrastructure, and may not be lowered efficiently directly through costly subsidies to some borrowers in the formal credit market.

4.2.2 Farm Size

As shown in Table 4.2, Thai rice farmers do not have equal access to credit sources. There appear to be significant differences in the distribution of institutional as well as non-institutional loans among farms of different sizes. Land holding size is positively related to borrowings, that is large farmers receive a greater proportion of loans from both sources, whereas a substantial portion of small farmers do not receive the credit they might need for either consumption or production. In the Northern and in the Northeastern provinces 63% and 45% respectively, of the smallest farms were indebted, while 80% to 100% of the largest ones had debts. Conversely, a higher proportion of non-borrowing farms in the largest holding size classes is observed in Suphan Buri.

It is likely that the non-borrowing large farms in Suphan Buri and Pitchit do not need to borrow at all because they belong to the richest regions and to the richest groups in the villages. These relatively wealthy households can smooth over income fluctuations with asset transactions; accumulation and decumulation of assets accomplish much of the requisite of expenses smoothing.

Small farmers' low participation in credit market is explained by other factors. A first explanation is simply that demand for credit is very low. Some of the small farmers' investment activities are too small to absorb large cash inflows, especially where no investment is being made in an improved technology. Then, borrowing costs might be disproportional relative to loans amounts for small scale borrowers. Finally, although all farmers are reluctant to incur land-secured debts if they perceive a high risk of losing their land through foreclosure, this is especially true for the poorest segment of farmers who fear that their ability to repay loans taken for investment purposes is very low. As a result, small farmers do not choose to borrow given existing terms.

Table 4.2: Distribution of Borrowers by Holding Size

	<15 rai	16-30 rai	31-50 rai	51-100 rai	101+ rai	All sizes
Suphan Buri						
Non-borrowing farms	25.0	11.9	9.1	66.7	100.0	18.0
Borrowing farms	75.0	88.1	90.9	33.3	0.0	82.0
from institutional sources	41.7	64.9	60.0	0.0		56.1
from non-institutional sources	41.7	16.2	20.0	0.0		24.4
from both sources	16.7	18.9	20.0	100.0		19.5
Pitchit						
Non-borrowing farms	36.8	28.0	17.2	20.0	20.0	24.5
Borrowing farms	63.2	72.0	82.8	80.0	80.0	75.5
from institutional sources	66.7	72.2	79.2	68.8	50.0	71.6
from non-institutional sources	33.3	11.1	8.3	0.0	25.0	12.2
from both sources	0.0	11.1	16.7	31.3	25.0	16.2
Roi Et						
Non-borrowing farms	55.0	27.3	25.0	0.0		42.7
Borrowing farms	45.0	72.7	75.0	100.0		57.3
from institutional sources	48.1	100.0	93.3	100.0		74.6
from non-institutional sources	44.4	0.0	0.0	0.0		20.3
from both sources	7.4	0.0	20.0	0.0		5.1

SOURCE: ORSTOM/CUSRI Project, Survey on Rice Farmers

However, another important reason that may reduce these households' need for borrowed funds, whether formal or informal, is the high frequency of off-farm employment which results in a substantial increase in internal funds (cf. chapter 2). In such a context, the demand for funds can be met from an inflow of funds from outside the farm sector. We must therefore consider the possibility that some small farms are less short of internal funds than is generally thought.

A second explanation rely on lenders' attitudes towards small borrowers. Small farmers are high risk borrowers who may not be considered creditworthy by credit institutions as well as by informal lenders. The case against lending to small-scale farmers is a straightforward one for a lender. It is expensive in terms of the higher risk of loans default as well as in terms of screening, monitoring and operating costs. Lenders, therefore, try to resist lending to small farmers, and concentrate more on large, less risky loans. In addition, small farmers' asset portfolio lack long-term tangible assets, especially land, useable as collateral. Collateral substitutes such as credit group, a share of the crop, reputation are subject to higher risk, particularly for those located in ecologically marginal areas, and are more difficult to enforce in case of default. As a result, small farmers often do not meet the eligibility criteria required by both formal and informal lenders.

As far as the structure of borrowings between sectors is concerned, there is evidence to show that farmers who have better access to institutional credit are among the richest in the rural community. The proportion of institutional clients tends to increase with the size class, while the opposite pattern prevails for non-institutional clients. Around half of the borrowing farms in the smallest size classes obtained their credit needs from institutional lenders, compared to 80% and more for the largest ones. Therefore, institutional lenders' operations have been oriented towards commercially viable medium and larger scale farmers and, for small farmers, informal credit is more readily accessible than formal credit.

The reasons for this relate to both supply and demand factors. On the supply side, formal lenders tend to discriminate against small borrowers because of costly information acquisition and weak enforcement capacity. Furthermore, the ceiling on interest rates imposed to institutions do not allow them to incorporate the high administrative costs that are involved in advancing small loans. Institutional lenders therefore employ various rationing mechanisms to steer away from what is seen as highly risky and costly loans[43]. These rationing practices involve allocating larger amounts of credit to carefully selected, economically secure farmers and result in market segmentation with adverse credit distribution consequences for small borrowers.

On the demand side, the extensive use of informal credit on the part of small farmers is often due to substantial borrowing costs associated with using formal sources. In practice, the borrowing cost includes two separate elements: the nominal interest payments and the transaction costs. Transaction costs involved in borrowing from the institutional sector are not cheap and present real barriers to access to institutional lenders (Adams and Nehman, 1979). Borrowers of small amounts and individuals who do not have borrowing experience with institutional lenders may incur relatively large transaction costs to acquire a loan. As a result, the real net cost of institutional loans for small scale borrowers might be much closer to the amount charged in interest by informal lenders than it appears at first glance. Accordingly, small borrowers might be « *indifferent* » as to whether to seek formal or informal loans because they feel their cost of acquiring a loan from either source is very similar.

Another powerful reason why many poor farmers actually prefer to borrow informally is that cash is readily available, and the terms of credit are more flexible. In contrast of institutional contracts many informal contracts are deliberately « *incomplete* », for instance by stating that the debtor will pay when he can. As cash flows vary in unpredictable ways, farms with insufficient assets find themselves unable to honor precise deadlines for payment. In such cases informal credit is a preferred source of finance for the farm business.

A very interesting fact which emerges from table 4.2 is that large farms tend to make greater use of multiple sources of credit than small ones. This tends to support the view that large potential borrowers are not able to borrow as much as they want at the prevailing interest rate, or, in other words, that they are rationed. For large farms, the amount of credit obtained from institutional sources has not been sufficient to cover production expenses and some money has to be borrowed from informal sources. The policy objective

---

[43] The use of non-interest charges to increase the real rate of return from loans have been reported by farmers during the survey. Substantial transaction costs, through the bothersome aspects of officialdom in particular, are imposed to small and new borrowers as a way of keeping unprofitable business away from the bank.

of helping an increasing number of farmers to take advantage of cheap credit had thus paradoxical results. Many creditworthy borrowers received too few funds from institutional agencies to capitalize on viable investment opportunities, resulting in a spillover of demand into the informal segment of the credit market.

The characteristics of the loan contracts are broken down by farm size in Table 4.3. The figures support first the conventional wisdom concerning the link between loan size and collateral. The size of the loan being limited by the value of the borrowers' collateral, there is a general tendency for larger scale farmers to secure larger scale loans, whatever the source. Comparing the top and the bottom groups in the Table 4.3, the average loan amount per transaction loans increases two-fold in Suphan Buri, four-fold in Roi Et and five-fold in Pitchit.

Second, since farm size (and thus assets size) is an indication of farmers' ability to bear risk, one would expect that small farms encounter difficulty in gaining access to long-term loans. Surprisingly, the term of the loan does not appear biased against small farmers except in Suphan Buri where the proportion of short-term loans varies from 70% for borrowers cultivating less than 15 rai to 50% for those who cultivate over 51 rai. In the other provinces, long-term loans are not concentrated among large farmers. The main reason for this apparent paradox comes from a probable overestimation of "real" long-term loans directed to small farms. Small farms are associated with a higher probability of default and, therefore, a higher conversion of short-term loans into long-term ones due to repayment problems.

Finally, though often hypothesized, there is little evidence from our data to support the assertion that the cost of borrowing declines with land size. Small sized farms do not bear higher interest rates than larger sized farms, once the source of the loan is controlled for.

Table 4.3: Distribution of Loan Characteristics by Holding Size and Loan Source

	<15 rai	16-30 rai	31-50 rai	51-100 rai	101+ rai	All sizes
Suphan Buri			*Institutional lenders*			
Loan amount (baht)	16,769	24,483	22,933	35,000		25,300
% of short-term loan	69.6	77.8	57.1	50.0		69.5
Interest rate	13.4	12.9	12.7	12.5		12.9
			Non-institutional lenders			
Loan amount (baht)	11,100	8,214	15,375	3,000		12,500
% of short-term loan	32.5	83.3	60.0	50.0		66.7
Interest rate	33.5	39.9	36.5	42.6		42.8
Pitchit			*Institutional lenders*			
Loan amount (baht)	14,750	23,620	25,143	32,143	80,000	32,700
% of short-term loan	58.3	58.8	73.1	72.2	66.7	67.1
Interest rate	12.7	13.3	13.2	12.1	14.5	12.9
			Non-institutional lenders			
Loan amount (baht)	16,500	15,500	25,000	13,250	20,000	19,700
% of short-term loan	75.0	25.0	80.0	57.1		76.9
Interest rate	48.4	42.6	54.3	37.2		43.2
Roi Et			*Institutional lenders*			
Loan amount (baht)	13,067	18,413	35,539		60,000	22,502
% of short-term loan	60.9	62.5	60.0			61.7
Interest rate	13.9	15.7	13.3	12.0		14.3

		Non-institutional lenders	
Loan amount (baht)	2,850	30,000	6,700
% of short-term loan	50.0	0.0	50.0
Interest rate	57.2		57.2

SOURCE: ORSTOM/CUSRI Project, Survey on Rice Farmers

4.2.3 The Security of Land Ownership and Land Tenure

Economic theory postulates that ownership security enhances incentives for investments of both capital and effort because the appropriation of the benefits from future returns on the investment is ensured. Ownership security is also expected to enhance capital formation by providing improved access to credit from both institutional and informal sources. Where suppliers insist on collateral, the credit market does not exist for groups or individuals who do not own assets which are acceptable as collateral or exist only at more adverse terms. Secure owners of land should thus be more apt to obtain credit especially from formal sources, provided that they are able to prove to the lender that a given and well-defined tract of land does indeed belong to them, and that they have the ability to transfer the land. Such an assurance is easy when a system of land registration and documentation exists.

The data pertaining to credit transactions of rice farmers according to land tenure security (as defined in Chapter 1) are given in Table 4.4. Available evidence suggests that partly secured owners were found to be the principal class beneficiaries of all sources of credit in each province: 95.8% of them reported credit transactions in Suphan Buri, 88.2% in Pitchit and 84.6% in Roi Et. In comparison, fully secure titled owners experienced apparently poor credit access.

Restrictive qualifications such as land titling tends to deny credit access, whatever the source, to untitled and tenants farmers only in Roi Et where 41.2% of untitled farms and 50% of tenants borrowed money, compared to 56.5% of fully secured owners and 84.6% of partly secure titled farms. In Suphan Buri and Pitchit, tenants face a disadvantage in the credit market as compared to titled and untitled owners, while farmers with unclear or no title do not seem significantly affected by credit rationing.

From the above, it would appear that ownership security by itself is not a major factor in explaining the restrictions cultivators may face when they apply for credit. Constraints based on land ownership security are more likely to affect households in remote and less developed areas. Obviously, the weight of ownership security on access to credit diminish in areas where output is largely commercialized and where higher income potential has been brought about by widespread adoption of new technology.

Table 4.4: Distribution of Borrowers According to Ownership Security

	Fully secure titled farms	Partly secure titled farms	Unsecure titled farms	Tenants
Suphan Buri				
Non-borrowing farms	29.2	4.2	-	12.0
Borrowing farms	70.8	95.8	100.0	88.0
from institutional sources	64.7	60.9	66.7	36.4
from non-institutional sources	23.5	21.7	33.3	27.3
from both sources	-	17.4	-	36.4
Pitchit				
Non-borrowing farms	29.6	11.8	15.4	28.6
Borrowing farms	70.4	88.2	84.6	64.3
from institutional sources	78.9	66.7	72.7	33.3
from non-institutional sources	5.3	13.3	18.2	22.2
from both sources	-	20.0	9.1	44.4
Roi Et				
Non-borrowing farms	43.5	15.4	58.8	50.0
Borrowing farms	56.5	84.6	41.2	50.0
from institutional sources	69.2	90.9	85.7	50.0
from non-institutional sources	28.2	-	-	50.0
from both sources	-	9.1	14.3	-

SOURCE: ORSTOM/CUSRI Project, Survey on Rice Farmers

Our results also point to the critical importance of the presence of a farm size effect. The persistence of pronounced size-related patterns of technique choice among farmers mitigates the effects of land title on production and investment incentives. The capital expenditure on farm, which is a function of farm size, has by and large a greater influence on borrowing than the tenure status. Partly secure farms who hold actually the largest farms, appear as the main credit recipient group.

To summary, there are different propensities among classes of cultivators to incur loans according to holding size and to the degree of development of agriculture. The credit operations of cultivators with large holdings, or cultivators located in relatively developed agricultural regions, whatever the tenure status, are well above the operations of the cultivators with relatively small holdings or the cultivators located in less developed regions. Lenders obviously consider differently farms with low levels of fixed assets living in relatively rich and well-endowed villages and farms living in resource-poor areas in which productivity is low and risky.

The relationship between ownership security and the use of institutional credit is not as clear-cut as expected. The proportion of partly or unsecured titled farmers who borrowed from institutional lenders is far from negligible: 66.7% of the untitled borrowing farms in Suphan Buri obtained their loans from institutional sources, 72.7% in Pitchit and 85.7% in Roi Et. The lack of full property rights in land is obviously not the main factor which prevents farmers from receiving institutional loans. Clearly, having a legal registered title is not a constraint for increased use of institutional credit. It may not be a necessary condition either, to the extent that suitable collateral substitutes exist.

However, differences in the incidence of borrowing from different sources are significant between owners (titled or untitled) and tenant farms. While 50% of the farmers who do not own the land they cultivate are institutional borrowers in Roi Et, no more than 36% of tenants in Suphan Buri and Pitchit are reported to be served by formal sources. Hence, tenant farmers called on informal sources of credit more often than owner farmers. This is not surprising since informal lenders are usually in a better position to lend without collateral; they are more able to select customers whose profile indicates that they have high repayment incentives. Enforcement is also less problematic because they can seize a wider variety of assets and more easily dispose of them than can a formal, non-resident lender. In addition, they can rely on social ostracism and coercion as effective sanctions.

The data presented in Table 4.5 show that as far as the loan characteristics are concerned, land security does not seem to be of great significance. Titled farmers have secured larger loans from both institutional and non-institutional lenders than unsecure titled and tenant farmers only in Suphan Buri. In the other two provinces, partly secure titled farmers received the largest loans. Another important observation is that loans obtained by tenants were, in many cases, larger than those obtained by unsecured farmers. This result is inconsistent with the conjecture that lenders decide on how much to lend based on land collateral provisions.

The percentage of households who received long term loans does not appear to vary greatly across tenure categories. However, the high percentage of long-term loans directed towards unsecure tenure categories, especially tenants, probably reflects difficulties in repayment performance leading to a conversion of short-term into long-term loans. Actually, it is very unlikely that this group enjoy easy access to long-term cheap credit to make investments such as the purchase of farm machinery.

Finally, there are no significant differences between titled and untitled farmers and between owners and tenants in the interest rates paid. By pledging land as collateral, the borrower does not obtain lower interest rates. The range of variation of average interest rates is therefore more pronounced between provinces than between classes of farms, and appears more influenced by differences in the liquidity of the rural economy as well as by risk differences in agricultural activity between provinces than by tenurial status.

Table 4.5: Distribution of Loan Characteristics According to Ownership Security

	Fully secure titled farms	Partly secure titled farms	Unsecure titled farms	Tenants
Suphan Buri		***Institutional lenders***		
Loan amount (baht)	32,600	19,900	10,000	21,313
% of short-term loan	35.5	37.0	34.0	23.8
Interest rate	12.7	13.6	13.5	12.6
		Non-institutional lenders		
Loan amount (baht)	18,300	9,833	3,000	10,250
% of short-term loan	22.2	35.3	40.0	28.6
Interest rate	31.8	43.4		37.0
Pitchit		***Institutional lenders***		
Loan amount (baht)	33,100	51,300	18,000	15,000
% of short-term loan	33.3	32.7	31.7	41.2
Interest rate	13.1	12.2	12.9	12.8
		Non-institutional lenders		
Loan amount (baht)	17,800	40,000	5,000	10,800
% of short-term loan	30.8	35.3	44.0	60.0
Interest rate	42.9	38.7	60.1	42.6
Roi Et		***Institutional lenders***		
Loan amount (baht)	22,800	24,417	14,100	30,000
% of short-term loan	38.2	34.1	37.0	50.0
Interest rate	14.4	13.8	14.1	17.0

		Non-institutional lenders		
Loan amount (baht)	4,900	30,000	10,000	1,000
% of short-term loan	44.4	50.0	54.5	100.0
Interest rate	56.3		79.6	42.6

SOURCE: ORSTOM/CUSRI Project, Survey on Rice Farmers

4.3 Access to Credit and Capital Formation

In theory, rural credit does have considerable potential as a catalyst to accelerate the adoption of new technology and thereby raising farmers' productivity and income levels. First, as mentioned by Binswanger and Khandker (1995), when liquidity constraints exist, additional credit supply can raise input use, investment and hence output (liquidity effect). Loans represent an additional source of liquidity which allow farmers to make an expenditure which may not have been possible ordinarily from existing resources and/or to make it in the proper time without drawing excessively on their own liquidity reserves. Second, better credit opportunities can help farmers to smooth out consumption and, therefore, increase the willingness of risk-adverse farmers to take risks and make agricultural investments (consumption smoothing effect).

4.3.1 The Puzzling Questions of «Attribution» and «Fungibility»

Methodologically, credit impact studies pose notoriously difficult problems to which this study is hardly immune. First, the evaluation of the impact of farm credit on investment in agriculture is complex owing to the number of factors other than the credit which affect the farmers' decisions. The links in the chain from loans to increased investment and thus productivity and incomes are subject to many disruptive influences, and uncontrolled variables may cause over- or underestimation. Isolating the effect of credit from other variables is complicated, and there is a risk of confusing the impact of credit with the effect of other factors that accompany it. There is thus the question of how much of the observed difference in capital use between borrowers and non-borrowers is attributable to credit, and how much to other uncontrollable factors such as differences in perception of risk, agroclimatic environment, or quality of irrigation and so on. This has been referred as the attribution problem.

Second, the link between the financial market and the rate of technical change in agriculture goes both ways. On the one hand, the availability of funds could determine the rate of innovation; on the other hand, the pace of technical change itself influences the demand for funds. Since progressive farmers with access to irrigation facilities and more advanced techniques are income to loans as it is from loans to these changes. The direction of causality is further obscured by the fact that farm capital represents both the long lasting process of physical asset accumulation and, as a store of wealth for the household, a base on which current credit could be allowed. This bias in credit access toward farmers with greater factors endowments implies that borrowers systematically differ from non-borrowers with borrowing the result rather than the cause. In other words, differences between farmers may explain credit allocation rather than the impact of borrowing. As a result, simple comparison of the performance of observed borrowers and non-borrowers is expected to overstate both the realized effects of credit on farmers who have obtained it and the potential effects on those who have not obtained credit. We thus need to disentangle the effects of credit from its causes.

Third, the households' liquidity reserves come from many sources such as crop income, off and non-farm income, rentals, loans and are then drained away to the household's most immediate needs at particular times of the year, so that neither the farmer, nor the researcher can confidently associate any particular source of liquidity with any particular expense. Since borrowed funds enter the household's total cash resources and become indistinguishable from other funds, it is usually not possible to say with any certainty which expenditures would not have occurred if any one source of liquidity, loans especially, had not been available. There are inevitably competing uses for cash received in the form of farm credit which may divert loans from their intended uses. This situation, known as the real expenditure substitution, where spenders treat the favored expenditure and the other expenditures as a common pool in the utilization of the credit, is prone to happen if the beneficiaries of the credit are not forced to utilize the credit on the expected expenditures. In such cases, loans may simply substitute for own savings or other sources of funds.

The problems of attribution, direction of causality and fungibility discussed above severely limit any attempt to measure the impact of credit on capital use in agriculture. In light of these problems, we did not attempt to make quantitative evaluations of credit impact on rice farmers investment in Thailand. Instead, we advocated the descriptive use of survey data to reach an understanding of the role of credit. Our approach depended on a comparison of borrowers and non-borrower farmers with respect to various measures of farming practices and input levels. The procedure was designed to control for potential effects of farm size and ecological conditions under which the farmers operate; differences in value of variables between borrowers and non-borrowers were thus broken into size and regional effect, so as to capture the contribution of each factor. This was expected to reduce attribution errors.

As far as fungibility is concerned, it has not been possible from the survey data to provide evidence of the extent to which loans obtained officially for production purposes were actually used as such. The declared purpose of credit was recorded during the survey. Interestingly, a significant proportion of loans were multi-purpose loans. In such cases, the exact proportion of credit which was intended for consumption, working capital or investment aims has been impossible to estimate[44].

Table 4.6 shows farmers' responses distribution concerning their intended use of formal and informal loans to finance working capital, investment in capital assets, family consumption needs or bad debts. Although it is often asserted that rural credit has little to do with production, few contracts were explicitly aimed at consumption expenses and debt repayment; actually, more than 80% of the loans were intended for production purposes. Even in the relatively stagnant agriculture of Northeast, many of the farmers who borrowed did so to finance rice farming activity outlays. However, much of the production credit meant to finance operating expenses such as labor costs and intermediate inputs; credit for long-term investment alone hardly approached 5% of total contracts.

[44]The distinction between consumption and working capital is sometimes misleading. To the peasant family farm, the amount of basic consumption can be an integral part of its operational costs. Peasants using credit for consumption needs should not be viewed as irrational and it is somewhat arbitrary to consider consumption expenses as unproductive. The head of household has to pay the minimum equivalent-wage to its family members in the form of food, consumption goods and so on. Borrowing in order to cover these expenses can be a necessary means to keeping the production unit running.

Table 4.6: Purpose of Loans by Source

	Institutional loans		Non-institutional loans		Total	
	Number	Percent	Number	Percent	Number	Percent
Suphan Buri						
Intermediate inputs	61	87.1	22	71.0	83	82.2
Capital assest	1	1.4	0	-	1	1.0
Family needs	7	10.0	8	25.8	15	14.9
Debt repayment	1	1.4	1	3.2	2	2.0
Total	70	100.0	31	100.0	101	100.0
Pitchit						
Intermediate inputs	61	93.8	16	80.0	77	90.6
Capital assest	1	1.5	1	5.0	2	2.4
Family needs	1	1.5	2	10.0	3	3.5
Debt repayment	2	3.1	1	5.0	3	3.5
Total	65	100.0	20	100.0	85	100.0
Roi Et						
Intermediate inputs	38	86.4	9	56.3	47	78.3
Capital assest	1	2.3	2	12.5	3	5.0
Family needs	5	11.4	5	31.3	10	16.7
Debt repayment		-		-		-
Total	44	100.0	16	100.0	60	100.0

SOURCE : ORSTOM/CUSRI Project, Survey on Rice Farmers

One should expect a higher share of non-institutional credit devoted to purposes not related to production for two main reasons. First, informal lenders (especially relatives and friends) are more inclined to lend money for purposes which are perceived as basic needs deserving assistance such as family needs and social events (funerals, weddings). Second, because informal resident lenders can monitor and regulate the borrower's behavior, they are likely to provide consumption loans more readily than can any other type of lender. Third, restraints on the use of loans from institutional lenders, especially denying their use for consumption purposes, are expected to push farmers to rely on informal credit for such expenses.

However, Table 4.6 reveals that, actually, informal credit has been available as a significant source of funding for farm investment in the study areas. Half of the loans for consumption needs are from non-institutional sources, while institutional lenders, mainly through cooperatives, provide the remaining half.

Theoretically, we should consider not only the purpose of different expenditures which are met through loan finance but also the level of these expenditures. Though the percentage of loans towards any expenditure may be large, the proportion of expenditure financed by loans might not be significant if the total aggregate expenditure is itself very high while the average loan amount is rather low. Assuming that loans are used as intended, Table 4.7 provides an estimation of the percentage of working capital expenses[45] financed out by short-term loans from both institutional and non-institutional lenders.

[45] These expenditures include expenses on fertilizers, pesticides, herbicides, seeds, fuel, and hired labor.

Table 4.7: The Percentage of Working Capital Expenses Covered by Loans

	< 15 rai	16-30 rai	31-50 rai	51-100 rai	101+ rai	All sizes
Suphan Buri						
Percent of working capital expenses covered by short term loans	27.5	24.3	12.5	1.0	-	19.0
Pitchit						
Percent of working capital expenses covered by short term loans	31.1	39.5	53.7	42.6	47.7	45.0
Roi Et						
Percent of working capital expenses covered by short term loans	82.6	62.3	82.7		-	72.0

SOURCE: ORSTOM/CUSRI Project, Survey on Rice Farmers

Inter-group differentials in this percentage are marked among provinces and within each province. In Suphan Buri, short-term loans tend to be limited compared with the expenses on working capital: about 19% of working capital has been met by short-term loans. In other words, farmers have depended on internal sources of funds to meet nearly 80% of the cost of material inputs and hired labor. In contrast, in Roi Et, about 70% of the cost of inputs has been covered by credit. This points to the existence of regional imbalances in the credit flow and a serious mismatch with the distribution of needs. Despite areas undergoing technological change have the highest absorptive capacity for credit, the present level of supply seems to be inadequate. Obviously, loans have not been critical to the operation of farmers in Suphan Buri and there is a strong presumption that the supply of funds has not met the rising needs for intermediate inputs.

As far as scale of operation is concerned, we know that small farms consume relatively more intermediate inputs than large ones; since their average income is much lower than those of large farms, they are expected to require relatively more borrowings to fund these expenses. Only in Suphan Buri are the ratio of borrowings of smaller farmers relative to their needs higher than those of larger farmers. In Pitchit, short-term credit covers a larger proportion of the total seasonal production costs of larger farms, while in Roi Et the size-effect appears insignificant. In these latter two provinces, small scale farmers appear as the group most likely to be subject to credit rationing.

4.3.2 Access to Credit and Farming Techniques

Table 4.8 gives useful insights into the characteristics of borrowers and non-borrowers and the farm level changes that can be associated with rural credit. Empirical evidence on the relationship between access to credit and farmers' production expenses and investment is mitigated. At the provincial level, the comparison of average inputs levels among borrowers and non-borrowers indicates that the expenses on intermediate inputs such as fertilizers, pesticides, herbicides, and hired labor are generally higher in the sub-group of borrowers. However, the differences observed are very small, indeed even insignificant.

Table 4.8: Per-Rai Inputs Values

	< 15 rai	16-30 rai	31-50 rai	51-100 rai	101+ rai	All sizes
Suphan Buri						
Intermediate input expenses (baht/rai)						
Borrowers	768	564	539	414		616
Non-borrowers	492	526	586	710	815	554
Hired Labor expenses (baht/rai)						
Borrowers	660	593	575			601
Non-borrowers	748	627	125	556	62	585
Value of agricultural eguipment (baht/rai)						
Borrowers	928	557	748	598		709
Non-borrowers	384	301	53	160	119	284
Pitchit						
Intermediate input expenses (baht/rai)						
Borrowers	327	312	162	212	310	244
Non-borrowers	326	296	207	141		250
Hired Labor expenses (baht/rai)						
Borrowers	181	433	341	406	385	354
Non-borrowers	328	395	197	239	143	298
Value of agricultural eguipment (baht/rai)						
Borrowers	869	810	496	332	514	598
Non-borrowers	1112	489	154	199	437	550

Roi Et						
Intermediate input expenses (baht/rai)						
Borrowers	182	154	127		68	158
Non-borrowers	209	156	144			195
Hired Labor expenses (baht/rai)						
Borrowers	136	213	207		324	177
Non-borrowers	115	169	206			133
Value of agricultural eguipment (baht/rai)						
Borrowers	19	166	372		376	155
Non-borrowers	75	20	3			59

SOURCE : ORSTOM/CUSRI Project, Survey on Rice Farmers

If we consider separate farm-size categories in each province, interesting patterns emerge. Operating expenses per rai are higher for borrowers who cultivate less than 30 rai compared to non-borrowers. This tends to confirm the proposition that some small farms are credit-constrained because they spend less on cash inputs as compared to small households who had access to additional liquidity through credit. However, as the scale of operation increase, the link between borrowing and expenses on intermediate inputs disappears. In the upper strata, borrowers are associated either with higher or lower use of modern intermediate inputs compared to non-borrowers.

Expenses on hired labor tend to be higher in the group of borrowers, whatever the scale of operations, suggesting that availability of credit had a stronger effect on labor use than on intermediate input use. This result is consistent with the fact that most households cultivate rice with traditional labor-using techniques. For such households, expenses on hired labor can represent up to 80% of total costs.

Long-term credit is generally expected to be closely associated with changes in the production relationship because long-term loans finance indivisible inputs more difficult to fund internally. Comparing borrowers and non-borrowers indicate that borrowers have significantly increased their acquisition of assets: there is a marked tendency for the per rai value of agricultural equipment to be the greatest in the borrowing farms, whatever the size of the farm and whatever the province considered. Hence, the possibility that some farm households are credit constrained provides one consistent explanation as to why farms with identical technical and environmental conditions, invest different amounts in agricultural machines.

On the whole, in view of the great increase in the adoption of highly divisible intermediate inputs such as fertilizers, pesticides and herbicides in irrigated areas and the relatively small differences in the expenses on these inputs among farmers whether they received credit or not, it is reasonable to conclude that adoption of this new technology in Thailand has not depended upon the availability of formal loans, except maybe for farmers holding less than 30 rai. By contrast, the provision of long-term credit appeared correlated with higher expenses on agricultural equipment. Consequently, the very low amounts of long-term credit distributed to rice farmers means that the mechanization process has probably been constrained by current loan supplies, and borrowing limits must be relaxed to obtain socially desired further changes in these variables.

4.4 CONCLUSION

The main aim of credit policy in Thailand has been to ensure cheap and plentiful credit to all rural households in order to encourage farmers' investments and to reduce their dependence on informal lenders. The dramatic extension in the availability of formal farm credit since 1975 has certainly represented a significant support for the rural sector. Rural banks have been under constant political pressure to expand their lending to farmers and succeeded quite well. Access to institutional credit has been easier and credit supply limits far more generous. The evidence presented in this Chapter shows that credit from the institutional sector has provided about 75% of total credit reported by rice farmers, compared with 25% for informal lending. Although it would be unrealistic to see credit relationships between farmers and private lenders as rare, alternative sources of institutional credit have become more accessible for a growing number of farmers and agricultural loans from formal financial institutions have been able to displace loans from informal sources among at least certain sub-groups of the population.

However, despite stated equity objectives, credit policy has been biased against small scale farmers. Institutional lending for agriculture is concentrated among wealthy farmers, who have a higher absorptive capacity for credit and a greater ability to repay. Small-scale ones, in comparison, have a lower probability of gaining access to institutional credit. Therefore, although the informal lender share in the total rural credit market has declined, their role among the rural poor has not declined accordingly. The survey tends to suggest that it is the greater access of well-endowed farmers to institutional credit which has mostly contributed to the general loss of the share of informal lenders.

That small-scale farmers are subject to institutional credit rationing is due to formal lenders' eligibility requirements structured to avoid adverse selection of borrowers and reduce the incidence of default. In addition, there are high administrative, operating and transaction costs on the part of financial institutions to deal with small borrowers, because a large number of small loans are expensive to process and supervise. Lending institutions such as

BAAC, while being allotted a developmental role by the government, are evaluated mainly in terms of repayment rates and profit on which financial viability depends. This contributed to the general resistance of public agencies to lending to small-scale farmers.

If it is clear that lack of ownership or of clear legal title prevents mortgaging the land, available evidence from the survey suggests that the impact of titling on access to credit is far less important than believed. Only titled farmers in remote areas enjoy a preferential access to any kind of credit. In other areas titled farmers do not borrow more frequently than untitled farmers. As regard the distribution of credit along with the source, we did not find any significant relationship between the possession of title and the use of formal credit. Moreover, title does not appear to imply an increase in loan maturity as it does for loan size. Thus, farmers with insecure ownership are disadvantaged mainly in terms of the loan size.

Of great significance for public policy is the interest rate the various lenders charge, for high interest rates have often been identified with exploitation and usury. However, the rice farmers' dependence on private moneylenders for credit appears, on average, of minor importance owing to the relatively low proportion of loans obtained from the informal sector. High interest rates are thus charged on only a small number of the debts incurred. Nevertheless, it remains that those who cannot borrow from institutional lenders are hurt by the differences in the rates and differential borrowing costs might strongly affect the willingness of the rural poor to seek for loans. Hence, the unequal distribution in the interest rate burden among farmers tend to reinforce the persistent income and productivity disparities.

Finally, the comparison of production expenses of farms that received credit with farms that did not shows that the provision of credit helped all farms to finance higher expenses in hired labor whereas it helped only small ones to purchase variable factors such as fertilizer. Expanded rural finance has also increased private investment in machines in all size groups. Higher lending volumes have thus led, on the whole, to modest increases in the use of intermediate inputs but substantial increases in both investments in equipment and agricultural employment.

CONCLUSION

The recent poor performance of the Thai economy has stimulated much speculation about the durability of the so-called Asian miracle. Although Thailand has clearly emerged from the financial and economic crisis which exploded in the mid-1997, the medium to long-term outlook remains uncertain. Some key macroeconomic indicators have stabilized (inflation, exchange rates, interests rates) and others have improved (current account surplus, foreign exchange reserves position). However, many factors are still hampering the economic recovery. Domestic purchasing power has collapsed, unemployment is mounting, production capacity utilization rates for most industries have fallen, net real investment has contracted, and the state of the banking sector is still precarious. The poor financial position of many private companies as well as the internal structural changes made by local banks in the name of improved risk management have led to the general reluctance of banks to make new loans.

Many questions for the nation have been raised, in particular how Thailand will continue to prosper. The pay-off to producing labor-intensive commodities relied for long on the country's surplus of labor at low opportunity cost, a condition that no longer prevailed by the end of the 1980s. Moreover, as neighboring countries were adopting similar strategies, they have been more and more able to out-compete Thai products. The Thai economic golden age of cheap-price advantage has passed. Therefore, Thailand's future economic growth depends, among other things, upon its ability to continually upgrade the technology and skills of its firms and workers.

The agricultural sector is one of the crucial pieces in Thailand's future growth as well as in the well-being of a majority of Thai people. Even though the overall export performance is poor, prospects for certain products offer some relief, food exports being, at the moment, the country's most sustainable export sector. The rice sector, in particular, could be supported as an economic buffer or a safety net. Thailand still has a comparative advantage as a rice exporter, in part because of natural resource endowments, which include soil and land quality, and because of Thailand's large past investments in rice production infrastructure. Rice is also the preferred cereal for most people in Asia, so demand prospects are good. As far as long-term changes are concerned, the FAO forecasts that if present trends in population growth continue, within 20 years, most Asian countries will no longer be self-sufficient in rice and Asia's legendary rice bowl could be filled increasingly by grain imports. To meet the rice demand over the next decades, it is estimated that average yields will need to reach about 6 tons/ha, nearly twice the current Thai level.

During the past decades, Thailand has experienced a rapid transformation of the structure of its economy. Consequently, rice households have been compelled to respond to stimuli from the external farm environment. Changes in external stimuli have been complex, including altered market trends for both output and inputs resulting from reallocation of production between sectors, changed agricultural policy measures, a deteriorated physical environment, the emergence of new competitors in the international market. Rice farm families adjustment strategies have included extension of the area cultivated, intensification of production, adoption of labor-saving technologies, development of other gainful activity both on and off the farm, and withdrawal from rice farming.

Agrarian Structure, Land Tenure Security and The Nature of the Land Market

Thai rice sector's spectacular performance in the past was accomplished mainly through the expansion of the land frontier. Strong population growth combined with easy access to land resulted in massive conversion of virgin land to agricultural use. According to an old Thai practice, every freeman had the right to occupy as much land as he and his family could cultivate. Land so opened and cultivated for a number of years later became the property of the occupier thanks to the implementation of numerous land laws, based on the principle that the rights to occupy or to use land constituted "property" rights.

This political choice had a crucial role in ensuring land ownership for an overwhelming majority of farmers. Furthermore, the ceiling imposed on farm holding size favored an agrarian structure dominated by small farmers. As a result, despite the significant population pressure that has occurred since the beginning of the twentieth century, most rice farms in Thailand are owned and operated by farm households.

The policy of *de facto* open access to land resulted in tremendous deforestation. The area under forest declined between 1.2 to 1.4 percent per year from the 1920s through the 1990s. Although the decision to convert forest into cultivated land depends basically on the profits from alternative land use, forestland-clearing process has doubtless been facilitated by lax enforcement of the existing legislation supposed to protect forest, the exemptions in the legislation itself, the exceptions accepted in practice, and the inefficiencies inherent in the administration of forestland. Encroachers were tolerated for a number of years and then suddenly noticed, rarely evicted, but given official status from time to time on a wide scale. The rate of deforestation leveled off recently mainly because the potential profitability of agriculture on currently forested lands has decreased. Low output prices, the development of higher wage non-agricultural jobs have been crucial in limiting forest clearing.

The impact of land tenure on farmers' efficiency is a recurring and important debate in economic theory. Economists have long regarded ownership as the single most important source of efficient resource allocation in the agricultural sector. Owner-cultivation is far more prevalent than any other form of tenure among rice farmers even since the closing land frontier in the 1980s. Landlessness and tenancy have not seriously developed because of huge amounts of unused land and, more recently, because "surplus" labor has been pulled to the urban labor market instead of being forced to compete for land ownership in the rural communities. Tenancy is more prevalent in highly productive areas whereas this tenure form is least common in rainfed areas. Mixed forms of tenure are common in Thailand so that the typical tenant is not a pure tenant but rather an owner-cultivator who expands his holding via contracting. As opposed to the view of tenancy as detrimental to agricultural development, land leasing appeared to provide an efficient allocation alternative to land transactions in the presence of multiple distortions in factor markets. The analysis of tenurial arrangements has shown that rented land was flowing from households lacking factors such as labor and capital to households better endowed with these factors.

Economic theory has recently deemed ownership security to be of fundamental importance in determining farmers' investments. The central argument of this supply side theory assumes a direct and beneficial link between private property rights and enhanced agricultural production and productivity. Secure ownership, it is argued, should ease land transfers, provide collateral for agricultural loans, and offer farmers incentives to invest in farm business.

Thailand has a long history of land titling and well-developed land laws. Increases in population density, commercial opportunity in agriculture, and political determination have driven the transformation of customary tenure, which granted only rights of use, into a complete set of property rights such as use, exclusion, and transfer. However, the lack of funds combined with the cost of the completion of a full cadastral survey and inadequate administrative infrastructure resulted in a very slow pace of title issuance until 1982. Since then, there has been a significant growth in the amount of land properly documented and in the issuance of the specific documents that are taken as evidence of legal ownership by banks.

It would be erroneous to assert that the efforts to enhance tenure security have worked as intended. Empirical evidence gathered during the survey suggests that the impact of titling on rice farmers' access to credit is on the whole negligible, especially in the richest areas of the country where the title's effect is overwhelmed by factors such as farm size. The issue of whether or not land can be used as collateral has proved to be irrelevant in a situation where small agents are denied access to formal credit markets because of the increasing screening and monitoring costs associated with the processing of a myriad of small and risky loans. Since the majority of rice farms tend to be small, private property rights in land offered little possibility of relaxing credit constraints on rice farming. Furthermore, the convincing success of small credit groups, which have been the principal channel for most of formal credit in Thailand, indicates that rural credit cooperatives and credit groups have a greater potential to contribute to meeting Thai farmers' demand for credit than land privatization.

Finally, the evidence on land transactions suggests that the land market is sluggish. One important reason for this is that the offer of sales in the market is limited by the unwillingness of farmers to part with land. Generally speaking, the more farmers were tied to the traditional and subsistence economy, the less land transactions were found. By contrast, land markets were found to be more dynamic in areas where land prices have risen due to increasing land productivity or change in land use (e.g. from rural into urban use). Systematic bias in land transfers from small-farm strata to large-farm strata tend gradually to alter the agrarian structure toward a more unequal pattern.

Industrialization, Labor Shortage and Adjustment in the Labor Market

The conventional view of the industrial revolution is that a process of structural change occurs in which both population and economic activities are urbanized. The analysis of Thai industrialization led observers to conclude that a slightly different process has been at work. Industrialization in Thailand has been characterized by a simultaneous movement of diversification of economic activity within the farm households rather than dramatic shifts of population from rural to urban activities. Development in the Kingdom has gone hand in hand with the expansion of seasonal employment opportunities for agricultural labor. As a result, the ratio of agricultural labor force still remains high (50 percent) and Thai farmers typically earn a large share of their total income from off-farm and non-farm sources.

Two forces leading farm families to diversify employment activities have been suggested in the literature, push factors (land shortage and poverty) and pull factors (increased opportunity). Our survey has identified a range of pressures that contribute to explaining why diversification into off-farm and non-farm work occurred. First, the local economy of rice farming has a strong effect upon household's off-farm earnings. Where agricultural productivity is low, and/or farm size is small relative to labor available, family labor is being employed outside agriculture because it has few opportunities to be productively employed in agriculture. Second, a young demographic structure usually indicates a high propensity to migrate, sometimes at the expense of further educational attainment of the household members concerned. Third, the accelerated labor intensive industrial development of the 1980s was accompanied by an increased demand for labor. Resulting widening wage gaps between agricultural and manufacturing sectors strongly pushed rural households to participate in outside labor markets. Fourth, standard explanations that emphasize distance to towns as main obstacles to labor market participation are not relevant for our sample of households, because the availability and quality of communication network brought distant areas quite close to main labor markets.

Since the 1980s, the rice labor force has been decreasing in absolute terms. Declining population growth rates, withdrawal of young and better-educated rural workers from rice farming in search of better paying jobs in the urban areas, and higher school enrollment have accounted for most of the decline in rice farming employment. The resulting shortage of labor has become particularly critical in irrigated areas because multiple cropping has led to sizeable increases in the need for both family and hired labor to cope with the bigger harvests and the higher intensity of management.

From a microeconomic point of view, the ability of any farm to bear the scarcity of labor is largely dependent on the scale of operation. Whereas the small farm sector appears structurally labor consuming, a larger farm size is generally associated with the choice of a more labor saving technology. Adoption of labor-saving technologies has been the technical solution that large farmers retained to face the rising need for labor.

Imperfect labor markets resulting in asymmetrical labor costs between large and small farms is an often-cited explanation for the differential in intensity of labor and fixed factor utilization. We reached the conclusion that imperfections in labor market cannot alone account for this inverse relationship. It has to be explained in terms of a multiplicity of market failures. Data support in particular the hypothesis of a credit constraint. Observed imperfections in credit market imply that the costs of mechanized technologies to small farmers differ widely from those relevant for large farmers. Therefore, unless specific credit lines are opened to small farmers for the purchase of farm equipment, the traditional small farm rice sector is the likely sector fated to disappear with labor scarcity.

The Process of Technical Innovation

Technical change is a dynamic two-stage process: first, appropriate technologies have to be generated, and second, they have to be implemented.

While the former stage usually involves public decisions about investment in research, the latter can be treated as a choice to be made by the farmer.

Technology adoption among rice farmers appeared heavily governed by the relative scarcities of land, labor, water and capital that farmers have to manage. Environmental factors such as the road network quality, quantities of government services available as well as public investment in irrigation also conditioned farmers' responses to new opportunities of investment. For given environment conditions and factor scarcities, the adoption of new technology further differed between the different groups of farmers because they are confronting different constraints due to their particular asset profile.

For a very long time, the Thai farmers means of coping with increased demand for rice output has been the expansion of the area under cultivation through the settlement of virgin land. The lack of a significant drive towards intensification is clearly reflected in the stagnant yields, the low fertilizer consumption in relation to international standards, the low cropping intensities, and the small proportion of area under high yielding varieties (HYVs). Many reasons are responsible for the persistence of this extensive path of growth. First, the potential benefits of intensive technology based on HYVs are largely dependent on reliable water control to secure and manage soil moisture. The large irrigation projects completed in many parts of the country have been disappointing regarding their limited ability to overcome the risks of drought. This low performance has mainly been due to insufficient water delivery and unproductive water losses. Lack of water in paddy fields has recently been aggravated because competition for scarce water resources has increased with industrialization and urbanization, while priorities have been biased towards the satisfaction of industrial and domestic water demand. Second, Thai people give considerable weight to varietal traits such as taste, and this is believed to have strongly affected the varietal composition of the output. Traditional crop cultivars such as glutinous or jasmine rice are typically low-yielding varieties which are not very sensitive to fertilizer application. Third, although there is some truth in the argument that unfavorable water supply conditions discouraged farmers from applying fertilizer, the heavy taxation imposed on rice farmers through the rice premium over a long period led to a very unfavorable rice-fertilizer price ratio. In addition, there has been no specific effort to bring new farmers into the group of fertilizer users. Fertilizer marketing has been handled mainly by a network of private distributors and therefore directed towards the commercially viable, " bankable " farmers who could either pay for fertilizer in cash or be sure to repay loans. As a result, the growth in fertilizer use has been concentrated on the agriculturally most advanced areas and farmers.

Greater investment in agricultural equipment was one of the major aspects of change that occurred in rice farming from the late 1960s onward. The pace of transition to mechanization has been partly driven by components of the growth process in other sectors that produced relative labor scarcity: a high level of private investment in the manufacturing sector and concentration in labor-intensive industries. The market for typical labor-saving machines such as harvesters and threshers expanded according to relative scarcity of labor, and thus rising wage rates. However, mechanization has not been induced only by relative price changes. The most important factor of water-pump investment has been the limited ability of the irrigation systems to provide an adequate water control. Adoption of small-scale irrigation technologies has thus been a key to increasing water use efficiency as well as a means to offset land scarcity. The shift from draft animals to tractors for power-intensive operations such as land preparation appeared also quite independent of labor costs. Small power tillers were adopted in paddy fields long before industrialization made labor scarce throughout the country. They were adopted to make possible the adoption of multiple cropping, when the time available for land preparation became very short.

Regionally, technical change has clearly not been neutral. The contrast between the rapid changes that occurred in Central Plain and their contemporary absence in most other areas of Thailand is particularly striking. The potential for agricultural development in the rainfed areas like Northeast has been very limited because of restrictions on climate, soil, and topography. Part of the differences observed between the North, the Northeast and the Central Plain in the intensification of rice production must therefore be attributed to the difference in bio-physical conditions in the three regions. But the availability of quasi-fixed public inputs such as rural roads, electrification, market infrastructure, and, overall, irrigation infrastructure has also been a crucial element in the diffusion of new techniques. Areas that received proportionately less government investment and services and/or areas with poor market connections had limited participation in these changes. Finally, the proximity of Bangkok to cities of Central Plain led to increases in the wages of agricultural workers and placed a high opportunity cost on the time of family members. This forced agriculture in this area to become more capital intensive so that productivity has been raised to cover the higher costs of labor and/or to allow family members to take temporary employment outside the agricultural sector.

Finally, the adoption of new technologies has not been either distributed evenly among different groups of individuals within given areas. While both large and small farms have successfully produced rice that met domestic and export demand, they have done so in different ways, with radically different factor combinations. The differences in the structure of labor and capital prices according to farm size have been the main determining factor of the contrasted demands for technological innovation. The investment in capital by large scale units has to a large extent been directed toward farm machinery, whereas small farmers along with greater labor input intensity are using intermediate inputs more intensively. Ineligibility for long-term institutional credit, which resulted in unfavorable terms of accessible (informal) credit, has prevented small farmers from proceeding with investments in farm machinery.

Performance of Agricultural Credit

Agricultural credit has long been a question of interest to policymakers as well as to development agencies. Up to the late 1980s, most governments of developing countries as well as donors set up directed credit programs aimed at channeling cheap resources to state-owned credit institutions for lending to farmers at below-market interest rates. A first objective was to provide an important ingredient to fuel agricultural growth and structural change. Another objective was to reduce the dependency of a certain class of borrowers on the informal credit market. Since 1975, the Thai government has placed increased emphasis on agricultural credit as a means to enhance rural development. Public and private institutions have been encouraged to provide increasing amounts of credit to agriculture. The resulting expansion of rural bank branches and increase in the volume of agricultural credit distributed have succeeded in bringing financial services to a wider segment of the rural population.

However, substantial gaps persist in the rural financial market. First, the creation of institutional alternative sources of credit has failed to drive the traditional moneylenders out of the market. Although institutional lenders are the leading source of farm credit, informal lenders still play an important role in serving farmers' needs. The simultaneous activity of many borrowers in both the formal and informal sector might suggest that formal and informal lenders are in direct competition. Our evidence tends to support the hypothesis of complementarity. The informal credit sector proved to have a strategic role to play, either by lending to those left out by the formal sector or for needs that the formal sector does not finance.

Second, there is evidence from the field investigations that larger farmers have had a disproportionate share of the available credit. Even so, rationing occurred since many of them were actually granted fewer funds than they wished to borrow at the going interest rate. Large farms require a broad set of financial services, and their credit needs were found to exceed the lending limit of institutional lenders, forcing these borrowers to resort to the informal sector for additional borrowing needs. A regulated increase in credit in favor of "priority" borrowers, i.e. small ones, has thus resulted in shrinkage of loanable funds for non-favored borrowers. Consequently, it is likely that many projects yielding relatively high returns have remained unfinanced.

Whether credit provision has been essential for the diffusion of innovation among rice farmers remains questionable. Credit is one of an inter-related complex of variables upon which output growth depends, including domestic agricultural terms of trade, input/output price ratios, public investment in rural infrastructure, and so on. Isolating the effects of loans from the effects of other changes in the farmer's environment has proved extremely difficult. However, it is likely that imperfections in the credit market have significantly contributed to the persistence of pronounced size-related patterns of technique and activity choice among rice farmers since both agricultural machinery ownership and access to institutional long-term credit is being biased towards the more creditworthy large farms. By contrast, on a per rai basis, the value of those types of current inputs such as fertilizer, pesticides are not higher on the borrowing farms than on the non-borrowing ones, which suggests that generally credit was not a binding constraint on crop-related investment. As for the rice farmers who exhibited unused borrowing capacity, their most likely constraint is an inadequate supply of investment opportunities that could generate demand for credit.

Under such conditions, giving loans to farmers at low interest rates does little to improve their position unless loans are accompanied by the provision of other inputs that could shift their production function as well as by a more favorable economic environment. On the whole, although the chances of using credit as a policy instrument to affect farmers' behavior appears to be great, a credit program alone may be unlikely to have much effect on innovation adoption or productivity. Increasing credit to agriculture should therefore be part of a wider set of measures.

SCOPE FOR GOVERNMENT INTERVENTION

During the 1990s, the changing environment characterized by the high cost of producing rice, the depressing trend of the real export price of rice, and the increasing competition for scarce resources among commodities and activities incited the Thai government to pursue a policy to replace rice with other crops. However, a recent analysis of the comparative advantage of rice and other competing crops concluded that agricultural diversification away from rice could result in an inefficiency loss (Yao, 1997). In addition, it may be possible that the increase in the price of cereals observed since 1995 signifies the halt or even the reversal of the long-term trend towards lower prices. Therefore, what policies should the government pursue for rice?

The Thai rice sector can no longer rely on cheap production factors because land resources have been exhausted, rapid industrialization has put the rural labor market under strong pressure while diverting more and more water from agriculture. Such a shift in the supply of crucial factors has clear implications for the categories of supporting investment that are needed for the future. If Thailand is to focus on the long-term viability of the rice sector and its comparative advantage in rice exports, increased emphasis on capital-led intensification is required to increase land and labor productivity as well as water use efficiency so as to face the country's rising factor scarcities and related factor costs.

Adequate water supplies are a critical condition of accelerated agricultural growth. Water policy will need to address a multitude of issues such as management of supplies (to improve water availability), management of demand (to improve efficiency of water use), and balancing competing demands (urban/rural; upstream/downstream). Whatever the policy options, reform of public behavior towards water will be a difficult task with high political and administrative costs. Disenchantment with the poor performance of large irrigation projects as well as the disappearance of most suitable sites for dams and reservoirs has focused the attention of policymakers on improved management of existing resources. There is a large scope for modernizing and making better use of existing irrigation. Rehabilitation can be achieved only at a fraction of the initial investment cost and could improve system efficiencies.

Debate on how to cope with inevitable future water scarcity has also led to the recommendation to recognize water as a scarce economic good and establish a water market so as to raise allocative efficiency (Christensen, 1994; Kaosa-ard and Kositrat, 1993). According to these authors, one of the ways to curb lavish water consumption would be to implement water fees which reflect the real value of water. However, charging farmers for water use is a sensitive topic since it will result in winners and losers. Moreover, security for a sufficient and reliable flow of water of suitable quality is a basic prerequisite for water pricing. The irrigation system needs to be improved before water rights can be tradable. The introduction of market-based policy instruments is therefore of a long-term nature.

Investment in biotechnology generation and diffusion should be accelerated to address critical issues such as resistance to pest and diseases, tolerance to drought, improvement of grain quality and response to use of chemical inputs,. Meanwhile, increased emphasis on efficient use of fertilizers will call for continuing action. The evidence accumulated through many studies and observations strongly suggest that market forces as well as government market actions such as heavy taxation of rice exports have seriously slowed down the process of fertilizer adoption. Supporting the private commercial supply system which is the backbone of the regular, country-wide fertilizer supply, with the elimination of the existing taxes (nearly 8%) on fertilizers could be an efficient and viable solution to reducing fertilizers cost. Also, more important than additional input use could be its more judicious application. While more evidence has to be provided, it seems that the current intensities and practices of fertilizer application are often technologically inefficient. Improving this efficiency through the provision of adequate extension services would help farmers to maximize expected profits.

Mechanization has a dramatic potential to increase labor productivity. In an economy such as Thailand in which structural transformation has occurred and the absolute size of the farm labor force has begun to decrease, investment in mechanization is likely to be highly profitable from society's point of view, and undoubtedly appropriate. The progressive substitution of mechanical for human and animal power will help to solve, in the long run, the problem of maintaining rice production with less labor. The government could provide the favorable conditions for a continuing development of the subsector of mechanization so that manufacturers and retailers are able to supply a large choice of machines at competitive prices. Further incentives for private machine producers in the form of tax exemptions on imported machine tools could be created[46], provided that the purpose and time limitations of these measures are clearly stated and understood. Another area for government intervention could be research and development. Future research and development in agricultural engineering could be more expensive or require skills that may not be affordable by private business, which has been the main investor in the past. Public research carried out in close co-operation with the private manufacturing sector would ensure that it is closely linked with the identification of markets. Finally, multi-use arrangements of existing machines through rental markets should be encouraged since they make available machines that farmers are not be able to afford individually.

Rural credit remains at a disadvantage with respect to industrial credit because of the dispersion of clients, the riskiness of agriculture and the typically large number of small loan accounts. Critics of the public agricultural credit programs have increasingly claimed that cheap institutional credit has been ineffective in stimulating agricultural investments and in spurring agricultural production. Indeed, the crucial issue in agricultural sector policy is the profitability of farming. As long as this remains low, lending for agricultural production is risky and inefficient. The expansion of the Thai rice production and exports was not driven by financial development, but by the availability of cheap land opened up by transport infrastructure and by a huge reservoir of manpower at low cost. The current problem of slow growth in the rice output is related to shortages of real factors like land, labor and water, not to the imperfection of the financial sector. An important policy implication of these considerations is that the generation and diffusion of improved investment opportunities could provide the government with a more powerful

[46] Local machinery makers must pay import taxes of up to 30 percent on machine tools, compared with duties of 3-5 percent on fully assembled imported machinery.

instrument with which to influence development in rural areas than formal credit.

Nevertheless, we argue that institutional medium- and long-term credit for investment in agricultural machinery is of a specific category because of the social profitability of this type of investment. As most small borrowers are short of high value collateral, the government could support the use of suitable forms of collateral substitutes in specific areas and for disadvantaged population groups. Due to the long duration of these types of loans, individual lending might be more appropriate than group lending. Loans used in purchasing fixed assets could be secured with mortgaged assets; the loan history of the farmer could be accepted as a collateral against loans; the amount of cash income derived from off-farm activities could be explicitly considered in eligibility for credit.

Lastly, in order to maintain their position in international markets or penetrate new ones, it will be necessary to produce good quality products, to innovate into new product lines so as to take advantage of market trends. This require skill-upgrading which may be the most serious challenge facing Thai industries. Investments in promoting better rice quality at all stages in the food chain could be quickly recovered through reduced losses, a better fit to consumer tastes, and higher prices. Although most of these investments may be expected from the private sector, the public authorities could develop and strengthen effective regulatory and quality systems for food control to meet international standards, invest in infrastructure to improve post-harvest handling, speed up the setting up of futures markets. As income will increase, changing demand patterns will result in a greater consumption of higher quality rice and processed products that demand less time for preparation. Another area of intervention could therefore be the support of joint cooperation between the government, researchers and private firms to develop processed rice products that have a higher value-added content.

REFERENCES

Adams, D.W., and Nehman, G.I., 1979. « Borrowing Costs and the Demand for Rural Credit », *Journal of Development Studies*, 15 (2): 165-76.

Alem, I., 1990. « Imperfect Information, Screening, and the Costs of Informal Lending: A Study of a Rural Credit Market in Pakistan », *The World Bank Economic Review,* (4) 3: 329-349.

Alexandratos, N., and de Haen, H., 1995. « World consumption of cereals: will it double by 2025? », F*ood Policy*, 20 (4): 359-366.

Baer, W., Miles, W.R., and Moran, A.B., 1999. « The end of the Asian Myth: Why were the Experts Fooled », W*orld Development*, 27 (10): 1735-1747.

Bardhan, P., 1979. « Wages and Unemployment in a Poor Agrarian Economy: a Theoretical and Empirical Analysis », *Journal of Political Economy,* (87): 479-500.

Bardhan, P. K., and Srinivasan, T. N., 1968. « Cropsharing Tenancy in Agriculture: a Theoretical and Empirical Analysis », *American Economic Review*, (51): 48-64.

Basu, K., 1986. « The Market for Land. An Analysis of Interim Transactions », *Journal of Development Economics*, (20): 163-177.

Basu, S., 1997. « Why Institutional Credit Agencies Are Reluctant to Lend to the Rural Poor: A Theoretical Analysis of the Indian Rural Credit Market », *World Development,* 25 (2): 267-280.

Bell, C., 1990. « Reforming Property Rights in Land and Tenancy », *World Bank Research Observer*, 5 (2): 143-66.

Bell, C., 1990. « Interactions between Institutional and Informal Credit Agencies in Rural India », *The World Bank Economic Review,* 4 (3): 297-327.

Berry, R.A, Cline W.R, 1979. *Agrarian Structure and Productivity in Developing Countries*, The Johns Hopkins University Press, Baltimore, 248 p.

Bertrand, T., and Squire, L., 1980. « The Relevance of the Dual Economy Model: A Case Study of Thailand », *Oxford Economic Papers*, (32): 491-92.

Biggs, T., and Others, 1990. « Rural Industry and Employment Study: A Synthesis Report, Rural Industries and Employment Project, TDRI, Bangkok, 131 p.

Binswanger, H., 1978. « The Economics of Tractors in South Asia: An Analytical Review », Hyderabad, ICRISAT, 96 p.

Binswanger, H., 1986. « Agricultural Mechanization: A Comparative Historical Perspective », *World Bank Research Observer*, 1 (1), pp. 27-56.

Binswanger, H., 1991. « Technological Change and Commercialization in Agriculture », The *World Bank Research Observer*, 6, (1), pp. 57-80.

Binswanger, H.P., and Deininger, K., 1997. « Explaining Agricultural and Agrarian Policies in Developing Countries », *Journal of Economic Literature*, 35: 1958-2005

Binswanger, H., and Khandker, S.R., 1995. « The Impact of Formal Finance on the Rural Economy of India », *The Journal of Development Studies*, 32, (2): 234-262

Binswanger, H.P., and Rosenzweig, M.R., 1984. *Contractual Arrangements, Employment and Wages in Rural Markets in Asia*, New Haven: Yale University Press, 330 p.

Binswanger, H.P., and Rosenzweig, M.R., 1986. « Behavioral and Material Determinants of Production Relation in Agriculture », *Journal of Development Studies*, 22 (3): 504-39.

Binswanger, H., and Ruttan, V.W., 1978. « Induced Innovation: Technology, Institutions and Development », Baltimore, The John Hopkins University Press.

Boserup, E., 1965. « The Conditions of Agricultural Growth », Chicago, Adine Publishing Company, 124 p.

Booth, A., and Sundrum, R.M., 1985. « Labor Absorption in Agriculture: Theoretical Analysis and Empirical Investigation », Oxford, Oxford University Press, 327 p.

Bot, K., 1983. « Mechanization and Employment in Thai Paddy Cultivation », ARTEP Working Paper, Bangkok, 80 p.

Braverman, A., and Srinivasan, T.N., 1981. « Credit and Sharecropping in Agrarian Societies », *Journal of Development Economics*, 9 (3): 289-312.

Byerlee, D., 1996. « Modern Varieties, Productivity and Sustainability: Recent Experience and Emerging Challenges », *World Development*, 24 (4): 697-718.

Charsombut, P., 1990. « Provincial Industry Labor Market », Rural Industries and Employment Project, TDRI, Bangkok, 57 p.

Christensen, S.R., and Boon-Long, A., 1994. « Institutional Problems in Thai Water Management », Working Paper, TDRI, Bangkok, 54 p.

Cuffaro, N., 1997. « Population Growth and Agriculture in Poor Countries: A Review of Theorical Issues and Empirical Evidence », *World Development*, 25 (7): 1151-1163.

David, C.C., and Meyer, R.L., 1980. « Measuring the Farm Level Impact of Agricultural Loans », John H. Ed., Borrowers & Lenders, Rural Financial Markets and Institutions in Developing Countries, London, Overseas Development Institute, pp. 201-35.

DORAS Project, 1996. « Agricultural and Irrigation Patterns in the Central Plain of Thailand », Kasetsart University, ORSTOM, 219 p.

Dyer, G., 1997. « Output per Acre and Size of Holding: The Logic of Peasant Agriculture under Semi-Feudalism », *The Journal of Peasant Studies*, 24 (1 & 2): 103-131.

Ellis, F., « Household Strategies and Rural Livelihood Diversification», *The Journal of Development Studies*, 35 (1): 1-38

FADINAP, 1984. « Marketing, Distribution and Use of Fertilizer in Thailand », ESCAP, Bangkok, 58 p.

Farrington, J., Abeyratne, F., and Gill G.J., 1982. *Farm Power and Employment in Asia: Performance and Prospects*, Bangkok, The Agricultural Development Council, 437 p.

Feder, G., 1983. « The Role of Family Labor, Supervision and Credit Constraints. », in The Relation between Farm Size and Farm Productivity, *Journal of Development Economic*, (18): 297-313.

Feder, G., and Akihiko Nishio, 1997. « The Benefits of Land Registration and Titling: Economic and Social Perspectives », *Land Use Policy*, 15 (1) : 25-43.

Feder, G., and Noronha R., 1987. « Land Rights Systems and Agricultural Development in Sub-Saharan Africa », *World Bank Research Observer*, 2 (2): 143-169.

Feder, G and Others, 1988. *Land Policies and Farm Productivity in Thailand*, World Bank, Baltimore, Johns Hopkins University Press, 165 p.

Feeney, D, 1982. The Political Economy of Productivity: Thai Agricultural Development, 1880-1975, University of British Columbia Press, London, 238 p.

Feeney, D., 1988. « Agricultural Expansion and Forest Depletion in Thailand 1900-1975. », in Tucker R., *World Deforestation in the Twentieth Century*, Duke University Press, London, pp. 112-287.

Fox, J., 1985. « Rural Credit Programs: The Case of Bank for Agriculture and Agricultural Co-Operatives BAAC. », Thailand, University of Hull, 133 p.

Fujimoto, A., 1996. « Rice Land Ownership and Tenancy Systems in Southeast Asia: Facts and Issues Based on Ten Village Studies », *The Developing Economies*, 24 (3): 281-515.

Grabowski, R., 1995. « Commercialization, Nonagricultural Production, Agricultural innovation, and Economic Development », *The Journal of Developing Areas,* (30): 41-62.

Griffin, K., 1982. *Land Concentration and Rural Poverty*, Holmes & Meier Publishers, New York, 338 p.

Hayami, Y., and Ruttan, V.W., 1970. « Factor Prices and Technical Change in Agricultural Development: the United States and Japan, 1880-1960 », *Journal of Political Economy*, 78: 1115-1141.

Heltberg, R., 1998. « Rural Market Imperfections and the Farm Size Productivity Relationship: Evidence from Pakistan», *World Development*, 26 (10) : 1807-1826.

Hoff, K., and Stiglitz, J.E., 1990. « Introduction: Imperfect Information and Rural Credit Markets-Puzzles and Policy Perspectives », *The World Bank Economic Review*, 4 (3): 235-250.

Hoff, K., and Stiglitz, J.E., 1997. « Moneylenders and Bankers: Price-Increasing Subsidies in a Monopolistically Competitive Market », *The Journal of Development Economics,* 52: 429-462.

Huppi, M., and Feder, G., 1990. « The Role of Groups and Credits Cooperatives in Rural Lending », *The World Bank Research Observer*, 5 (2): 187-204.

Hymer, S., and Resnick, S., 1969. « A Model of an Agrarian Economy with Non-Agricultural Activities », *American Economic Review*, 59 (4): 493-506.

Ingram, J., 1971. *Economic Change in Thailand 1850-1970*, Stanford University Press, Stanford, 352 p.

Inukai, I., 1970. « Farm Mechanization, Output, and Labor Input: A Case Study in Thailand », *International Labor Review*, (101): 471-72.

Ishii, Y., 1978. « Thailand: A Rice Growing Society », Monographs of The Center for Southeast Asian Studies, The University Press of Hawaii, Honolulu, 340 p.

Isvilanonda, S., and Poapongsakorn, N., 1995. « Rice Supply and Demand in Thailand: The Future Outlook », TDRI, Bangkok, 133 p.

Jain, S., 1999. « Symbiosis vs. Crowding-out: the Interaction of Formal and Informal Credit Markets in Developing Countries», *Journal of Development Economics*, (59): 419-444.

Johnston, B.F., and Cownie, J., 1969. « The Seed-Fertilizer Revolution and Labor Force Absorption », *American Economic Review*, (58): 569-582.

Kemp, J., 1981. « Legal and Informal Land Tenures in Thailand », *Modern Asia Studies*, 15 (1): 1-23.

Kaosa-Ard, M., and Kositrat, N., 1993. « Economic Instruments for Water Resource Management in Thailand, TDRI, Bangkok, 21 p.

Khan, M. M., « Economic Status of Households and Labor Supply to Market and Non-Market Activities - Some Results From a Poor Rural Economy », *Applied Economics,* (27): 779-791.

Larson, D., and Mundlak, Y., 1997. « On The Intersectoral Migration of Agricultural Labor », *Economic Development and Cultural Change*, 295-313.

Lasker, B., 1950. *Human Bondage in Southeast Asia*, University of North Carolina Press, 406 p.

Lewis, W.A., 1954, « Economic Development with Unlimited Supply of Labor », *The Manchester School of Economic and Social Studies*, 22 (2): 139-191.

Lynch, D., 1996, « Measuring Financial Sector Development: A Study of Selected Asia-Pacific Countries », *The Developing Economies*, 34 (1): 3-33.

Long, M.G., 1986. « Why Peasant Farmers Borrow », *American Journal of Agricultural Economics*, 50 (4): 991-1008.

Manarungsan, S., 1989. « Economic Development of Thailand 1850-1950 », Institute of Asian Studies, Chulalongkorn University, Bangkok, 265 p.

Mehl, C., 1986. « Social and Cultural Aspects of Land Transactions in Rural Thailand. », Agricultural Research Unit Discussion Paper World Bank, Washington D.C., 52 p.

Mehl, C., 1985. « Land, Agricultural Labor, and Development in Rural Thailand: A Study of the Variations in the Socio-Economic Structure of Three Villages in Nakorn Ratchasima Province. », Ph.D. Dissertation, Cornell University, 281 p.

Mukherjee, A., and Ray, D., 1995. « Labor Tying », *Journal of Development Economics*, 47: 207-239.

Onchan, T., 1982. « The Burden of Debt of Thai Farmers: An Overview. », Kasetsart University, Bangkok, 32 p.

Onchan, T., 1984. « Off-Farm Employment and Rural Development. », Staff Paper 52, Department of Agricultural Economics, Faculty of Economics and Business Administration, Kasetsart University, Bangkok, 47 p.

Onchan, T., 1985. « Farm Mechanization Policy in Thailand. », Bangkok, Center for Applied Economics Research, Kasetsart University, 25 p.

Onchan, T., 1985. « Credit Problems of Farmers in Thailand », *Farm Credit Situation in Asia*, Tokyo, Asian Productivity Organization, pp. 51-71.

Onchan, T., 1990. « A Land Policy Study », Research Monograph 3, Bangkok, TDRI, 136 p.

Panayotou, T., and Parasuk, C., 1990. « Land and Forest: Projecting Demand and Managing Encroachment », TDRI Year-End Conference, Bangkok, 85 p.

Pant, C., 1983. « Tenancy and Family Resources: A Model and some Empirical Analysis », *Journal of Development Studies*, 12: 27-39.

Phananiramai, M., and Tonguthai, P., 1994. « Labor Welfare in Thailand », Final Report, Bangkok, TDRI, 65 p.

Phelinas, P., 1995. « Croissance et Compétitivité de la Riziculture Thaïlandaise », Communication Présentée au Colloque CNRS/CIRAD sur l'Avenir des Rizicultures en Afrique de L'Ouest, Bordeaux, 22 p.

Piker, S., 1976. « The Closing of the Frontier », *Contribution to Asian Studies*, 9: 7-26.

Platteau, J.P. , 1995. « A Framework for the Analysis of Evolving Patron-Client Ties in Agrarian Economies », *World Development*, 23 (5): 767-786.

Platteau, J.P., 1996. « The Evolutionalry Theory of Land Rights as Applied to Sub-Saharan Africa: A Critical Assessment », *Development and Change*, 27 (1): 29-86.

Pokpong, S., 1985. « Factors Affecting the Household Indebtedness: A Case Study in Nangrong District. », Bangkok, Institute for Population and Social Research, Mahidol University, 31 p.

Rabibhadana, A., 1969. « The Organization of Thai Society in the Early Bangkok Period 1782-1873 », Data Paper N° 74, Department of Asian Studies, Cornell University, New York, 247 p.

Ramsay, A., 1985. « Population Pressure, Mechanization, and Landlessness in Central Thailand », *The Journal of Developing Areas*, 19 (3): 351-68.

Rattanabirabongse, V., 1998. « The Thailand Land Titling Projetc: Thirteen Years of Experience », *Land Use Policy*, 15 (1): 3-23.

Rauniyar, G.P., and Goode, F.M., 1996. « Managing Green Revolution Technology: An Analysis of a Differential Practice Combination in Swaziland », *Economic Development and Cultural Change*, 44 (2): 413-437.

Reinhardt, N., 2000. « Back to Basics in Malaysia and Thailand: the Role of Resource-Based Exports in Their Export-Led Growth», *World Development*, 28 (1): 57-77.

Rosegrant, W.M., and Herdt, R.W., 1981. « Simulating the Impacts of Credit Policy and Fertilizer Subsidy on Central Luzon Rice Farms, The Philippines », *American Journal of Agricultural Economics*, 63 (4): 655-665.

Rosenzweig, M.R., and Wolpin, K.I., 1985. « Specific Experience, Household Structure and Intergenerational Transfers: Farm Family Land and Labor Arrangements in Developing Countries », *Quarterly Journal of Economics*, 961-987.

Ruttan, V.W., 1997. « Induced Innovation, Evolutionary Theory and Path Dependence: Sources of Theorical Change », *The Economic Journal*, 107 (444): 1520-1529.

Sadoulet, E., Fukui, S., and De Janvry, A., 1994, « Efficient Share Tenancy Contracts under Risk: The Case of Two Rice-Growing Villages in Thailand, *Journal of Development Economics,* (45): 225-243.

Sen, A.K., 1964. « Size of Holdings and Productivity », *Economic and Political Weekly*, (111): 323-326.

Sen, A.K., 1975. « Employment, Technology and Development », Oxford, Clarendon Press.

Setboonsarng, S., Werakarnjanapongs, P., 1988. « Forestry and Land Use Policy », Bangkok, TDRI, 50 p.

Shand, R.T., and Others, 1986. « Off-Farm Employment in the Development of Rural Asia », National Center for Development Studies, Canberra, The Australian National University, 367 p.

Siamwalla, A., and Others, 1990. « Thai Agriculture: Resources, Institutions and Policies. », Bangkok, TDRI, 51 p.

Siamwalla, A., 1991. « Land-Abundant Agricultural Growth and some of its Consequences: The Case of Thailand. », Bangkok, TDRI, 53 p.

Siamwalla, A., and Nettayarak, P., 1989. « Estimating the Cost of Subsidies for Agricultural Credit. », Thailand Development Research Institute Foundation, Bangkok, 14 p.

Siamwalla, A., and Others, 1990. « The Thai Rural Credit System: Public Subsidies, Private Information, and Segmented Markets », *The World Bank Economic Review*, 4 (3): 271-295.

Siamwalla, A., 1991. « Rural Credit and Rural Poverty », Thailand Development Research Institute Foundation, Bangkok, 37 p.

Singh, I., 1988. « Small Farmers in South Asia: Their Characteristics, Productivity, and Efficiency. », World Bank Discussion Papers, Washington, D.C., 75 p.

Singh, I., 1988. « Land and Labor in South Asia. », Washington, D.C., The World Bank, 236 p.

Singh, R.D., 1996. « Female Agricultural Workers' Wages, Male-Female Wage Differentials, and Agricultural Growth in a Developing Country », *Economic Development and Cultural Change*, 89-123.

Skoufias, E., 1995. « Households Resources, Transaction Costs, and Adjustment Through Land Tenancy, *Land Economics*, 71 (1): 42-56.

Stiglitz, J. E., 1990. « Peer Monitoring and Credit Markets », *The World Bank Economic Review*, 4 (3): 351-366.

Stiglitz, J. E., and Weiss, A., 1981, « Credit Rationing in Markets with Imperfect Information », *The American Economic Review*, 71(3): 393-410.

Stifel, D.L., 1976. « Patterns of Land Ownership in Central Thailand During the Twentieth Century », *Journal of The Siam Society*, 64 (1): 237-74.

Sukharomana, S., 1982. « The Impact of Farm Power Strategy in Thailand », Proceedings of a Regional Seminar, Colombo, Agrarian Research and Training Institute, pp. 139-59.

Sussangkarn, C., 1987. « The Thai Labor Market: A Study of Seasonality and Segmentation », Macroeconomic Research Conference Paper n° 9, Bangkok, TDRI, 56 p.

Sussangkarn, C., 1990. « Labor Market in an Era of Adjustment: A Study of Thailand », Bangkok, TDRI, 22 p.

Sussangkarn, C., 1991. « Thai Economic Growth, Emerging Labor Market Problems and Policy Responses », Bangkok, TDRI, 31 p.

Sussangkarn, C., and Chalamwong, Y., 1994. » Development Strategies and Their Impacts on Labor Market and Migration: Thai Case Study », Bangkok, TDRI, 47 p.

Sussangkarn, C., 1998. « Thailand's Debt Crisis and Economic Outlook », Bangkok, TDRI, 9 p.

Tinakorn, P. and Sussangkarn, C., 1996. « Productivity Growth in Thailand », Research Monograph n°15, Bangkok, TDRI, 235 p.

TDRI, 1990. « Policy on Agricultural Land Reform in Thailand », Bangkok, 106 p.

TDRI, 1992. « Food Situation Outlook in Asia: Case Study of Thailand », Bangkok, TDRI,
173 p.

Tohtong, C., « Joint Liability Groups for Small Farmer Credit: The BAAC Experience in Thailand », *Rural Development in Practice*, (1): pp. 4-7.

Tongpan, S., and Others, 1990, « Deforestation and Poverty: Can Commercial and Social Forestry Break the Vicious Circle? », The TDRI Year-End Conference, Chon Buri, Thailand, 176 p.

Tower, I., 1997. « Mexico 1994 versus Thailand 1997 », *TDRI Quarterly Review*, 12 (3): 9-14.

Vajragupta, Y., and Vichyanond, P., 1998. « Thailand's Financial Evolution and the 1997 Crisis », Bangkok, TDRI, 47 p.

Wade, R., 1995. « The Ecological Basis of Irrigation Institutions: East and South Asia », *World Development*, 23 (12): 2041-2049.

Wattanutchariya, P., 1983. « Economic Analysis of the Farm Machinery Industry and Tractor Contractor Business in Thailand. », in Swaminathan My.S. Ed., Consequences of Small-Farm Mechanization, IRRI, Los Banos, Philippines, pp. 39-49.

Wenner, M. D., « Group Credit: A Means to Improve Information Transfer and Loan Repayment Performance », *The Journal of Development Studies*, 32 (2): 263-281.

World Bank, 1980. « Thailand: Case Study of Agricultural Input and Output Pricing », World Bank Staff Working Papers, n° 385, Washington, D.C., 134 p.

World Bank, 1983. *Growth and Employment in Rural Thailand*, Country Programs Department, East Asia and Pacific Regional Office, Washington D.C, 157 p.

World Bank, 1985. « Thailand: Pricing and Marketing Policy for Intensification of Rice Agriculture », World Bank Country Study, Washington, D.C., 164 p.

Yao, S., 1997. « Comparative Advantages and Crop Diversification: A Policy Analysis Matrix for Thai Agriculture », *Journal of Agricultural Economics*, 48 (2): 211-222.

Yao, S., 1997. « Rice production in Thailand seen through a policy analysis matrix », *Food Policy*, 22 (6): 547-560.

INDEX

G

H

I

J

L

M

N

O

P

T

U

V

W

Y